DATE DUE	RETURNED

This book is part of the Peter Lang Media and Communication list.
Every volume is peer reviewed and meets
the highest quality standards for content and production.

PETER LANG
New York • Bern • Frankfurt • Berlin
Brussels • Vienna • Oxford • Warsaw

Joanna L. Jenkins

The Convergence Crisis

AN IMPENDING PARADIGM SHIFT IN ADVERTISING

PETER LANG
New York • Bern • Frankfurt • Berlin
Brussels • Vienna • Oxford • Warsaw

HUMBER LIBRARIES LAKESHORE CAMPUS
3199 Lakeshore Blvd West
TORONTO, ON. M8V 1K8

Library of Congress Cataloging-in-Publication Data
Jenkins, Joanna L.
The convergence crisis: an impending paradigm shift in advertising /
Joanna L. Jenkins.
pages cm
Includes bibliographical references and index.
1. Advertising. 2. Mass media—Technological innovations.
3. Digital media. I. Title.
HF5823.J456   659.1—dc23   2014042551
ISBN 978-1-4331-2607-9 (hardcover)
ISBN 978-1-4331-2606-2 (paperback)
ISBN 978-1-4539-1489-2 (e-book)

Bibliographic information published by **Die Deutsche Nationalbibliothek**.
Die Deutsche Nationalbibliothek lists this publication in the "Deutsche
Nationalbibliografie"; detailed bibliographic data are available
on the Internet at http://dnb.d-nb.de/.

Cover design by Joanna L. Jenkins

The paper in this book meets the guidelines for permanence and durability
of the Committee on Production Guidelines for Book Longevity
of the Council of Library Resources.

© 2015 Joanna L. Jenkins
Peter Lang Publishing, Inc., New York
29 Broadway, 18th floor, New York, NY 10006
www.peterlang.com

All rights reserved.
Reprint or reproduction, even partially, in all forms such as microfilm,
xerography, microfiche, microcard, and offset strictly prohibited.

Printed in the United States of America

TABLE OF CONTENTS

Acknowledgments	vii
The Convergence Crisis: Introduction	1

**Part I: The Road to Normalcy:
The Development of American Advertising, 1840s–Present**

Chapter 1:	Early Beginnings	11
	Early Beginnings	13
	Visionaries	13
	Factions	18
	Tumultuous Environment	21
Chapter 2:	Dramatic Differences	29
	A Turning Point	31
	A Unifying Paradigm	33
	Interdisciplinary Collaboration	34
	Strategic Use of Modernity	36
Chapter 3:	Growth & Expansion	41
	Cultural Fusion	42

	Technology & Turmoil	44
	War & Peace	46
	Colossal Expansion	47
	Shifts & Upheavals	51
	Hills & Valleys	57
Chapter 4:	Drift & Decline	67
	Influence & Obligations	69
	Creative Oasis	71
	Groundbreaking Precedent	74
	Cultural Currency	81
Chapter 5:	Paradox & Pitfalls	95
	Democratized Information	97
	Triadic Convergence & Radical Change	100
	Associations & Accountability	104

Part II: Triadic Convergence, Insights, and Implications

Chapter 6:	Overlapping Phenomena	125
	Triadic Convergence	125
	Advertising	129
	Overlapping Phenomena	132
	Insights	134
Chapter 7:	Triadic Convergence & The New Media Ecosystem	139
	Themes Associated with Triadic Convergence	142
Chapter 8:	Moving Forward	153
	Advertising in Crisis	153
	Paradigm Shift	154
	Triadic Convergence: Support & Criticism	159
	Moving Forward	163
	Appendices	169
	References	177
	Index	189

ACKNOWLEDGMENTS

In acknowledgement to the source from which all is possible, thank you.

This endeavor has been realized with the support of many. My list of names and specifics is without end. Thank you all!

In recognition of the gracious support of colleagues and mentors: Dr. Barbara Hines, Dr. Tony McEachern, Dr. Clint Wilson II, Mark Bartley, Dr. Marcia Y. Cantarella, Dr. Rochelle Ford, Dr. Emory Tolbert, and Dr. Gwendolyn Bethea.

In recognition of the sustaining prescience and encouragement of family and friends: Leonard J. Jenkins Sr., Sharon Jenkins, Jesse L. Jenkins, Debra Jenkins-Kearney, Abrielle M. Kearney, Zurie S. Kearney, Ted Kearney, Leonard "Buck" Jenkins Jr., Lydia Francis-Jenkins, Nailah Jenkins, Elijah Jenkins, George Jenkins, Joseph Jenkins Sr., Rebe Jenkins, the Jenkins family, Richard Carter Sr., Richard Carter Jr., Carol Blalock, Regina Smith, the Smith-Carter family, Theodore Wing III, Joi Jackson, Tyear Middleton, Deana Diggs, Christina Johnston-Brown, Alisha Cowen-Vieira, Maya Oliver, Richelle Johnson, Dr. Crystal Adkisson, Dr. Bradford D. Wilson, and Dr. Adria Goldman.

THE CONVERGENCE CRISIS
INTRODUCTION

Recent years of advertising have been an accelerated path of calamity. Yes, there has been success. Yes, there has been progress, but, overall, advertising is not winning. There have been major—no, excuse me—epic fails! I'm sorry, is that still cool? That's still cool, right … that's not passé. That's not cliché. It's not like I said "winning"; that's old. But if I said "winning" in the right context it would be OK … right? There are periods where everything just seems to run together in a contradictory mess. But there is hope. There is social media—the great fixer. Social media will not only rectify all advertising woes but it will also bring the world together in a very sincere Kumbaya moment. And that's it. Problem solved. But, just when you think you have it together—boom. Something else happens, such as Instagram, "the gram," Pinterest, Vine, Buzz Feed, or the next best thing ever. Don't worry, this is life. This is advertising. Everything is under control. This is how the pendulum swings. Things happen and everything will get back to normal soon. But no, something major has happened. Something has drastically altered reality. There has been a huge shift. There are now multiple voices in this conversation and no one is quite sure who is in control or even if control is still an option. One advertising professional stated, "I walk around in fear and loathing, dazed and confused. I feel like I'm standing here and there are a thousand baseballs dropping from the sky and I don't know which ones to catch" (Sacks, 2010). If this could be you; if you can relate to this at all; if you are involved in mass communications. If you are a member of the advertising community—an advertising professional, an advertiser, a practitioner, researcher, educator, or student, and you sometimes find yourself unable to understand what is going on, welcome to the convergence crisis.

The convergence crisis is a paradoxical juncture that represents an opportunity to redefine one of the world's most powerful institutions. What seems to be the end of advertising could actually be its new beginning. Although the convergence crisis has been accompanied by seemingly unmatched turmoil, it offers avenues for empowerment and the potential for extraordinary achievement.

During this period advertising has been filled with confusion, contradiction, and chaos. Although there has been significant progress, there have also been setbacks and false starts. In the 1980s advertising began to drift toward decline. Advertising expenditures peaked in 1984 and have trended downward ever since (Rust & Oliver, 1994). Recent decades have revealed a traversed landscape in which the old and new comingle in intricate and nonconventional ways. In the new media ecosystem, digital technology has revolutionized communication, earned media has usurped paid media, and consumers dictate brand conversation. Although digital advertising is lucrative, it is not growing nearly fast enough to keep pace with declines in legacy ad formats, such as print and television (Holcomb & Mitchell, 2014).

Normative structures, formidable systems, and established relationships have been disrupted, leading to economic turmoil. Massive layoffs, restructuring, and monopolies have characterized the industry. Global conglomerates, including the Interpublic Group, the WPP Group, the Omnicom Group, and Publicis Group, have acquired formidable advertising agencies (Inukonda & Pereira, 2010), while other independent agencies and creative boutiques have collapsed under pressure.

Transformative patterns in demographics, production, and consumption have also created pressure for advertising. Target audiences have become savvy, migratory communities with increased demands and expectations. Advertising has struggled to determine effective methods to engage modern consumers. Many of these methods have increased confusion as consumers can no longer easily decipher advertising messaging amid media clutter, covert content, and unfamiliar formats.

Simply put, advertising is in crisis and there is no conclusive delineation for its extent. Moreover, deteriorating conditions are expected to worsen due to advertising's inability to effectively measure, monetize, and create revenue.

The primary intention of this book is to demystify elements surrounding the convergence crisis and to increase awareness regarding the opportunities it presents. At its core, it answers two main questions: How did we get here? and What do we do now? A discussion of how a state of crisis erupted in contemporary American advertising and suggestions on how to navigate in this complex new environment are offered.

A fresh perspective presented. It poses a new way to engage with advertising and its development that goes beyond biographic and anecdotal information. The underlying intention of this book is to advance scholarship and to assist in the restoration of pragmatic activity within the advertising community. Readers will discover points of intersection for increased collaboration among research, practical application, and academia such as

- exploring patterns of conflict and resolution
- learning how advertising traditionally responds to new technology, cultural shifts, and economic recession
- uncovering dominant power groups
- sanctioning knowledge proprietors
- establishing relationships within advertising.

This information is particularly useful when projecting the future of advertising.

In support of its framework, *The Convergence Crisis* is organized into two parts. Each part can be read consecutively or independently. Part 1 (How did we get here?), "The Road to Normalcy: The Development of American Advertising, 1840s–Present," explains how a state of crisis erupted in contemporary American advertising. A historical chronology traces the development of the discipline of advertising, its collision with convergence, and its path to crisis. Due to advertising's broad influence, part 1 includes interconnected elements of American business, popular culture, burgeoning technology, and the ever-changing American public.

Part 2 (What do we do now?), "Triadic Convergence, Insights, & Implications," discusses major themes surrounding the impact of convergence on advertising. Ways to harness convergence and navigate within a complex media environment are shared. Additionally, part 2 offers insights and practical tools for empowerment.

To aid this discussion, specific conceptualizations of convergence and advertising have been offered for clarity. Both convergence and advertising are complex phenomena. Over the years, they have been subjected to extensive usage, major shifts, and broad assertions. Although each will be explored in depth, it is important that underlying conceptualizations are well established.

Triadic Convergence

Convergence is a source of modern controversy. It has revolutionized nearly every aspect of daily life. In a broad sweep, this powerful force has altered everything from the global economy and industry to the environment and projections of the future. Convergence has blurred boundaries and caused a redistribution of power, influence, and functionality. These are often precursors to radical change, growth, and innovation.

Due to widespread implications, conceptualizations of convergence are often general and abstract. This book offers a more specific meaning of convergence as it relates specifically to advertising and institutions. Decidedly, a conceptualization of convergence more closely aligned with advertising may lead to deeper understanding, problem solving, and solutions.

To assist with this endeavor, triadic convergence—a complex and dynamic force comprised of the sophisticated intermingling of three core elements: media, technology, and culture—is offered. Intermingling is characterized by the constant mutative and adaptive synergy achieved among media, technology, and culture. This interwoven relationship is shaped and shifts in accordance with the characteristics of the shifting locus of power within the triad.

Figure I.1. Triadic Convergence Model.

For increased understanding, triadic convergence is accompanied by a model. The triadic convergence model is a departure from previous models of convergence, which depicted three or four circles gradually moving toward each other (Brand, 1987; Wirtz, 2001). The circles within previous models typically represented information technology, telecommunications, media, and consumer electronics. These four industries were expected to merge, resulting in eliminated barriers and American industry functioning as a seamless entity.

The model for triadic convergence specifically addresses complexity. It depicts three united moving circles representing a singular yet dynamic force. Each circle is dimensional and represents a locus of power within the triad. No circle rests distinctly on an axis. Rather, each circle overlaps and intersects in a symbolic representation of the complex intermingling of media, technology, and culture.

In this depiction of triadic convergence, media is positioned at the top. This positioning represents centralized power within contemporary advertising characterized by widespread conglomeration. Technology and culture are positioned opposite each other at the base of the model. This positioning is symbolic of the continuous relationship between technology and culture in which the cause and effect are sometimes reciprocal and not fixated. The model also includes dimensional arrows, which represent the erratic nature and mutative intermingling inherent to convergence. It is important to note that the positioning of the three elements, media, technology and culture, is not permanent and will shift as change occurs.

Triadic convergence is accompanied by several suggestions for interpretation. Although these suggestions will be explored later in this discussion, the most relevant asserts that major advances in electronification spur convergence. Electronification is the control of the flow of electrons that facilitates pulses of electromagnetic energy to embody and convey messages. When observing convergence within advertising, it is important to keep in mind that advances in technology have a direct impact on media. Most common, advances in technology drive economic processes of cross-ownership (de Sola Pool, 1983).

In a normal environment, media monopolies are largely determined by the capabilities of technology. Increased competition is the standard in digital environments, which are highly characterized by technological deregulation. Conversely, monopolies are prevalent in digital environments that are heavily regulated. As mentioned, contemporary advertising is characterized by monopolies and conglomeration. However, contemporary advertising is characterized by technological deregulation as well. Due to the collision of these elements, an unstable environment has erupted.

Advertising

Now that a meaning of convergence has been established, attention will be directed toward advertising. In this book advertising is conceptualized as an institution to best demonstrate how it impacts society as a whole. Institutions are implemented to uphold authority, create structure, and preserve order at micro- and macro levels. Advertising commands significant influence over media, individuals, cultures, purchasing habits, and economies worldwide. Conceptualizations of advertising that are limited in scope may diminish its collective impact.

Framing advertising as an institution is attributed to mass communications scholar James Carey. Vincent Norris furthered this concept with the addition of criterion. Figure I.2 depicts essential characteristics associated with advertising as an institution. It is important to emphasize that advertising is a fixed yet continuously evolving phenomenon. As a result, advertising's responses to shifts are rarely fluid or quick.

A dimensional triangle is used in the depictive model of advertising. Each point is representative of a characteristic of institutional advertising. Point 1,

Figure I.2. Institutional Advertising Model.

which rests at the top of the model, represents advertising's ability to order human relationships into roles. Two points support the base of the model. Point 2 represents advertising's ubiquity. Point 3 point represents advertising's ability to regulate the distribution of society's essential resources.

When implementing an institutional view of advertising there are several claims to consider. The first claim states that the source of advertising is determined by the demands of technology and the location of economic power (Carey, 1960). When placed within the context of contemporary advertising, this claim raises immediate concerns as there are conflicts surrounding technology and economic power.

Recent conflicts have largely stemmed from modern technology's ability to disperse power among diverse audiences. For well over half a century, power was disseminated through a fixed linear structure that flowed from advertisers to audiences. In the past, power structures primarily benefited traditional producers, such as advertisers, clients, advertising agencies, and conglomerations. Accordingly, many of advertising's economic practices, standards, and relationships were based upon these fixed structures. Within contemporary advertising, boundaries surrounding these structures have begun to disintegrate. As a result, power flows in multiple directions among advertisers and audiences.

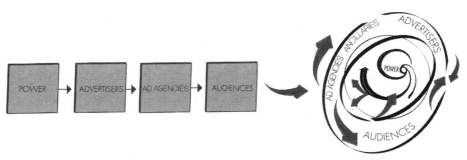

Figure I.3. Power Disruption Model.

The impact of triadic convergence on advertising could be revolutionary. Over the years research has shown that developments in social responsibility, technology, and power structures have a great impact on advertising. Shifts in these areas affect advertising's function and the role advertising plays within society and the economy. Within contemporary advertising these conditions may be even more intense as many pertinent issues have been exacerbated and interlaced. Present shifts can lead to a redefinition of advertising as well as consumer behavior and expectations.

Due to its enormous power and influence, changes in advertising will significantly impact nearly everything, from the global environment and economy to American ethos and society. The present malleable state of advertising is extremely important. Within this plasticity, valuable contributions can be made. As society sits on the cusp of new beginnings, advertising is pregnant with possibilities. The future will be shaped by the choices that are made now.

This discussion is of widespread critical importance. A state of crisis and prolonged transition has been entered with no definitive end in sight. Massive confusion and the need for understanding are abounding. Although this paradoxical juncture presents a plethora of opportunities for extraordinary achievement, it also represents the dissolution of many well-established traditions. Clarity and navigation are paramount.

To attain these goals, the historic development of advertising, its collision with convergence, and the path to crisis will be traced. This will assist in gleaning tools for navigation in our complex and contradictory environment. Remnants of the past contain insights for the present as well as the outlines for the future. Particular attention will be placed upon major events that influenced social-cultural conditions, the economy, technological uptake, and historic foundations.

The journey begins in the 1840s alongside the development of advertising and a wave of production associated with the Industrial Revolution. From there we travel to contemporary times symbolized by the Internet phenomena during the 1990s and the social media revolution of the late 2000s. Along the way we will examine major conflicts, such as World War I and the Vietnam War, and how these events impacted advertising. Cultural benchmarks, such as the civil rights movement and individualism, are investigated to determine their effect on modern consumerism. Pop culture milestones, such as the birth of MTV and the influence of hip-hop culture, are included to demonstrate advertising's seamless interconnectivity to media and entertainment. Technological breakthroughs, highlighted through Adobe and post-script language, reveal how technology and electronification transform institutional industry and business practices.

The development of advertising is remarkable. It is an invigorating drama filled with human desire, passion, entrepreneurialism, intellect, and contradiction. Above all, it is an insightful and inspirational journey characterized by promise, triumph, failure, and new beginnings. Without further ado, let's begin our travels.

PART I
THE ROAD TO NORMALCY: THE DEVELOPMENT OF AMERICAN ADVERTISING, 1840S—PRESENT

· 1 ·

EARLY BEGINNINGS

In the nineteenth century, American advertising developed amid the hustle and bustle of a changing world. Fervor swept the nation as technological advances, economic productivity, and migration propelled America in new directions. Simultaneously, factions and turmoil threatened to disrupt its core. Although advertising was shaped by the efforts of visionaries and entrepreneurs, it was also heavily influenced by numerous environmental factors.

Advertising's immediate environment swelled with an array of unprecedented issues and concerns. America expanded its borders into territories previously held by diverse and international cultures. The primary interests of the nation began to shift as millions of acres and thousands of communities now became one. Urbanization and industrialization led to new social, political, and economic challenges. Slavery and contrasting ideologies marked brewing conflicts. It was during these times that advertising emerged as a fledgling discipline from a kaleidoscope of progress and reform.

Advertising developed within three distinct phases. As a framework for this historical chronology a brief overview of each phase is provided. The first phase of development, early beginnings, began in the 1840s and lasted for nearly 70 years. During this phase advertising agents worked vigorously and provided the groundwork for future productivity.

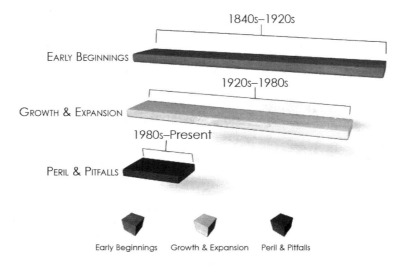

Figure 1.1. Development of American Advertising 1840s–Present.

Visionaries and entrepreneurs seized opportunities that came about as a result of economic and industrial productivity. However, knowledge was highly concentrated, which contributed to factions and competing interests. Although there was progress, diffusion of knowledge had yet to occur. As a result, advertising struggled to gain the momentum necessary to move forward collectively.

The second phase, growth and expansion, began in the 1920s shortly after advertising attained widespread knowledge and acclaim. In what proved to be a critical turning point, members of the advertising community were heavily recruited to join the George Creel Committee on Public Information. This committee, which became known as the world's greatest adventure in advertising, was commissioned by President Woodrow Wilson in 1917 to aid World War I.

Advertising made tremendous strides as a result of its involvement with the Creel Committee. The number of factions decreased and advertising matured. Knowledge was diffused throughout the discipline as new technologies and methods were embraced. Subsequently, a consensus was reached and a collective approach to advertising was selected. The newly established paradigm allowed the discipline to gain momentum and prosper. Advertising forged relationships with powerful entities and remained productive for decades.

The 1980s marked perils and pitfalls as advertising began to drift toward a state of crisis. After advertising reached an apex, it began to exhibit signs of serious concern. New and empowering technologies emerged that contributed to the disintegration of industry structures and renewed factions within advertising. Once profitable relationships mutated and became difficult to manage. Furthermore, previous methods of communication were unable to sustain profit and influence consumer behavior.

After the continual inability to sustain profits, engage consumers, and harness technology, productivity ceased throughout advertising. The universal standstill signaled that a state of crisis had begun. Moreover, the cessation of productivity was a stark indication that the paradigm used to govern advertising was now grossly insufficient.

Long after the world changed, advertising remained deeply entrenched in its traditional practices. Major phenomena, such as convergence, and technological advancements transformed the world and audience expectations. Advertising struggled to gain footing in an ever-changing environment. By the mid-2000s advertising reached a crossroads filled with critical decisions, confusion, and angst.

Although there has been some progress, the need for clarification still remains. Lucidity exists within the unfolding of advertising's history. A deeper understanding of the present is gained from starting at the beginning. Within rich details of development, insights are revealed.

Early Beginnings

The early beginnings of advertising, 1840s–1910s, were characterized by pioneering visionaries, competing factions, and a tumultuous environment. Like many budding institutions during this time, advertising was young, formless, and under the influence. The spirit of entrepreneurialism and the promise of opportunity drove trailblazers to create paths that were not readily apparent. However, competing opinions, public disdain, and multiple vested interests inhibited collective progression. Nonetheless, the early beginnings of advertising provided the foundation for future growth and expansion.

Visionaries

The impact of advertising pioneers revolutionized business practices and helped create new methods of operation. The efforts of one visionary, Volney Palmer, became legendary due to his enormous success and range of influence. Palmer opened the first advertising agency in Philadelphia, Pennsylvania, during the 1840s. His client roster was unprecedented and unparalleled. Palmer's clients spanned the East Coast, Midwest, and Southern regions of the United States. By the close of the decade, Palmer represented more than half of the media vehicles during his time.

Much of Palmer's success was attributed to his thorough understanding of advertising. He had a keen ability to convey how advertising could be used to create profits for business. Palmer's persuasive tactics were reinforced by his sound business practices and methods of communication. Palmer instituted systematic operations and innovative ways to target consumers and reach large audiences. Notably, Palmer provided advertisers with space rates and commissions systems, which increased efficiency and customer satisfaction.

Palmer's innovation extended well beyond systematic methods of operation. His efforts helped to spearhead the emergence of advertising agents, which revolutionized advertising during the 1840s. Prior to the emergence of advertising agencies, advertising practitioners worked primarily for media. Advertising practitioners considered themselves agents for newspapers and other media publishers rather than advertisers. Practitioners' primary duties included soliciting orders, transmitting copy, collecting payments, and selling space for media publishing (Fox, 1997).

Although gradual, the emergence of advertising agents created a significant impact within industry. In expanded roles, advertising agents began to provide counsel on fiscal budgets, offer impartial advice to advertisers, and assume central responsibilities in the construction of advertising (Fox, 1997). Palmer, in particular, began to craft advertising copy. This act affirmed the growing authority of advertising agencies. Previously, the creation of ads was the responsibility of media publishers. Advertising agents also served as knowledgeable liaisons between media and advertisers.

Undoubtedly, advertising agents and agencies represented significant shifts in power and expansion. As the relationship between media and advertisers diminished, the advertising industry increased in power and influence. Advertising seized opportunities that emerged as the nation became a more industrial society. Advancements in transportation and steam engine productivity expedited business processes and encouraged further development.

Advertising leveraged beneficial conditions. As a result, the number of advertising agents and agencies increased. In regards to Palmer, successful conditions facilitated his ability to open three additional branches of his agency. By the 1850s, Palmer had offices in Boston, New York, and Baltimore. In the 1870s, additional advertising visionaries emerged. James Walter Thompson opened his self-titled advertising agency and invented the role and responsibilities of account executives. Albert Lasker, a revered and highly influential leader in advertising, filled this position and later ascended to the agency's presidency.

Despite advances, knowledge of advertising and client rosters were not widespread. Although there was some progress, the concentration of knowledge and advertisers crippled collective achievement.

Visionaries who became successful in advertising created strong relationships with clients, in addition to conducting sound business practices. Such information was not readily shared throughout the industry. Although growing, advertising was still a small and relatively young industry. Many who advanced were reluctant to share information with competition. Moreover, advertisers were not frequently willing to conduct business with lesser-known and inefficient agencies or advertising agents.

In many cases, knowledge and clients remained fixed within the same power blocs for decades. For example, by 1860, Palmer's agency became Volney B. Palmer & Company upon the establishment of a partnership with John E. Joy, J. E. Coe, and W. W. Sharpe. After Palmer's death in 1864, Volney B. Palmer & Company became Coe, Wetherill & Company and was subsequently incorporated into N. W. Ayer & Son in 1877. The Ayer Agency rose to prominence, popularizing the use of slogans and successful economic practices. In 1898, the Ayer Agency launched national advertisements for the first prepackaged biscuit, Uneeda, utilizing the slogan "Lest you forget, we say it yet, Uneeda Biscuit." Although agencies varied over the years, later incarnations benefited greatly from the rich history, solid reputations, loyal clients, and systematic operations of previous agencies.

Client loyalties were further strengthened by negative perceptions of advertising held by the public. Many Americans believed advertising was quackery. Advertising agents were despised and perceived as untrustworthy. Advertising was frequently associated with the introduction of inferior products to the market. As stated by advertising historian Steven Fox (1984):

> Advertising was considered an embarrassment—the retarded child, the wastrel relative, the unruly servant kept backstairs and never allowed into the front parlor. A firm risked its credit rating by advertising; banks might take it as a confession of financial weakness. Everyone deplored advertising. (p. 15)

Thus, an advertising agent that was credible was often lauded by his clients and rewarded with repeat business.

Credible advertising agents stood out for two primary reasons. American advertising was young and therefore void of government regulation and ethical standards. Second, the early beginnings of advertising gained momentum during the Gilded Age (1870s–1900s), an extremely tumultuous period of American history. Despite rapid economic growth, advances in communication technology, and industrial productivity, abject poverty, corruption, and social upheaval were widespread. Controversies erupted as America struggled to find solutions to many of its most pressing issues, which included Prohibition, labor unions, human rights, and suffrage. Consequently, fervent reform movements swept the nation and greatly influenced the ethos during that time.

Advertising developed during a period of intense reflection on morals and ethics. In addition to pioneers who advocated their progressive vision, reform efforts shaped advertising as well. Visionaries believed advertising had to gain credibility in order to mature as a discipline and retain clients. Advertising leaders also contended that public approval would contribute to increased business and revenue. Palmer and other advertising leaders, notably Samuel Pettengill and George P. Rowell, pushed advertising in directions that would complement their vision (Fox, 1997).

Although dedicated, advertising visionaries faced an uphill battle. Without government regulation and sufficient ethical standards, advertising became riddled with corruption and inconsistency. Systems of operation resulted in profuse variations, pricing scandals, and covert arrangements. Additionally, advertising agents had a dubious public reputation. Agents did not maintain clear and transparent communication. Agents often exaggerated or lied about the circulation and reach of media publishers (Fox, 1997). Moreover, it was customary for an agent to purchase space from media vendors at the lowest price and then sell the same space to advertisers for the largest profit possible. Such practices resulted in strained relationships with media and advertisers.

Assiduously, advertising began to make significant progress. Formidable achievements included standardized circulation statements, the use of contracts, and major campaigns substantiated by national advertisers. George P. Rowell championed much of the reform process surrounding media standardization. After a number of local improvements—including a survey of newspapers, special contracts and rate systems—Rowell offered his most significant contribution. In 1869 Rowell premiered *Rowell's American Newspaper Directory*, which included circulation figures as a choice feature. Within this directory Rowell accessed the integrity of national newspapers and provided advertisers and advertising agents with critical information and vital data.

While Rowell's efforts helped eliminate many of advertising's issues with media, Francis Wayland Ayer worked diligently to improve advertising ethics. Ayer implemented a contractual system that was widely adopted throughout the industry. Rather than acquiring payments through purchasing at the lowest costs and selling for the largest profit, agents now adopted a set commission system. Agents received between 12% and 15% commission based on media publisher's fees. Moreover, Ayer's contractual system bound agents with advertisers through trusted financial terms and order. Agents now worked directly for advertisers, in contrast to previous times when agents worked for themselves or for media publishers. Ayer's system eliminated a major source of corruption within advertising and established ethical credibility for agents.

Advertising credibility was further strengthened with the success of national merchandise. Success was largely achieved through simplified distribution methods and product improvements. Advances in transit, communication systems, and manufacturing capabilities created new distribution channels. Previous barriers between manufacturers and consumers were eliminated as the power of idiosyncratic local markets was diminished. Advertising assumed more significant roles and functions as it became a viable tool to provide information about goods and services.

Unlike patent medicines, which were frequently affiliated with advertising, newly manufactured goods and services were reliable, convenient, and safe. Thus, the validity of advertising claims improved as a result of the trustworthy nature of these products. Joseph C. Hoagland, a visionary who was both an advertising agent and inventor, leveraged this unique situation and achieved exponential success. Hoagland invented and marketed Royal Baking Powder, a premixed substitute for yeast. Hoagland placed advertisements for his product in well-established religious and women's periodicals. He only patronized periodicals that used commission systems. Hoagland was also among the first to feature an image of his product in its advertisements. Moreover, Hoagland used the tagline Always Pure and selected premiere positioning for ad placement, including the back cover. By the early 1890s Hoagland attained one of the largest budgets in advertising, $600,000 annually.

In addition to the success of Royal Baking Powder, products such as Ivory Soap, Douglas Shoes, and Sapolio soap contributed credibility to advertising. Each of these products was driven by visionaries who leveraged opportunities that emerged within a changing marketplace. Artemas Ward, advertising manager of Enoch Morgan's Sons soap-making firm, worked for more than two decades to ensure that Sapolio became a widely recognized brand. Ward used unprecedented incentives and tactics when compared to traditional advertising standards. In addition to the use of bold out of home and guerilla tactics, Ward circulated a mysterious legend that created intrigue among the public.

Much like the success of national products, visionaries drove the early beginnings of advertising. Although there were several noteworthy contributors, Palmer, Rowell, and Ayer are distinct. Their efforts significantly shaped the birth of modern advertising in America. Volney Palmer broke ground with the opening of the first advertising agency. Through leveraging favorable conditions and applying sound business methods, Palmer created powerful new roles and functions for advertising. George P. Rowell contributed to this vision and the credibility of advertising, by smoothing relationships and defining responsibilities. Rowell's efforts helped to standardize practices and expectations between advertising and media publishers. Francis Wayland Ayer furthered the credibility of advertising through implementing contracts and systematic operations. Ayer's diligence resulted in the decline of corruption within advertising and the improved perception of advertising among the public and advertisers.

Although visionaries helped to provide a strong foundation for advertising, a great deal of work still needed to be accomplished. Despite improved perceptions, advertising was plagued by public skepticism and disapproval. In addition to external factors, advertising struggled through internal conflicts as well. Neither success nor knowledge was pervasive within advertising. Polarizing disparities, rivalries, and harsh competition contributed to the emergence of factions. With sharp divides it was difficult to sustain the momentum needed for collective progression.

Factions

Throughout its early beginnings there were several advertising factions. These advertising factions included, but were not limited to, retail advertising, market-driven advertising, agency development, and atmosphere advertising. In addition to factions, ancillary groups, which possessed their own unique opinions, existed as well. Due to its infancy, advertising was malleable and subject to broad interpretation.

Varied opinions, incompatible ideology, and incomplete rationales were widespread, which led to dissention in advertising. Each faction embraced distinct beliefs and core values. Their beliefs greatly influenced what role they believed advertising should have in society as well as what functions it should provide. Factions were motivated by their specific agendas. Each jockeyed for the dominant position (Fox, 1997). Often, key spokesmen from factions behaved as advocates for their groups' ideals among broader audiences. As stakeholders, if the future of advertising affirmed their views, they would attain positions of power, leadership, and authority.

Despite passionate attempts, convincing evidence that would produce an accord among the majority of the advertising community had yet to emerge. This was no easy task. In achieving consensus, factions would have to abandon their core beliefs and ideologies. With widespread discontent and bitter rivalries, consensus seemed highly unlikely. Although there was progress toward growth and achievement, knowledge remained insufficient. Advertising could establish neither a collective agenda nor a cohesive direction. Factions coexisted in a "snipping stalemate" for a large duration of advertising's early beginnings (Fox, 1984, p. 74).

One ardent faction within advertising was retail. This faction advocated a strategic approach to advertising that was coined "ads as news" (Moriarty, Mitchell, & Wells, 2012, p. 13). Under this approach, advertisements resembled classified listings. Retail ads were constructed with an abundance of copy and very little imagery. Retail advertising was primarily backed by department stores, which have a rich and influential history in advertising.

One of the greatest influences in retail was John Wanamaker and his self-titled Philadelphia department store. With fixed prices and generous refund policies, Wanamaker's attracted a large audience and substantial revenue. Wanamaker became an avid advertiser and frequently used handbills, billboards, window displays, and newspapers to communicate with consumers. In 1880, Wanamaker established yet another retail precedent. Wanamaker hired the first full-time, in-house advertising copywriter in retail, John E. Powers. This influential adman gained prominence among retailers through his work with Lord & Taylor department store (Fox, 1984). Powers made formidable contributions to Wanamaker's and retail advertising. Not only did he popularize informational body copy and simple language construction, Powers influenced branding, content, and tone.

Powers led many retail advocates to believe that the primary objective of advertisements was to identify products and to deliver pertinent information (Pope, 2007).

In contrast, market-driven advertising was primarily concerned with consumer behavior rather than informational needs (Fox, 1997). Advertising legend Albert Lasker, of J. Walter Thompson, championed this approach with the phrase "salesmanship in print driven by the reason why." The implementation of this method required that advertisements state a claim followed by an explanation of product support (Moriarty et al.). Proponents believed this method would motivate purchasing behavior among audiences.

While some placed focus on consumers, other advertising factions emphasized the importance of agency development. Advocates of this approach believed that attaining professionalism and strong client relationships provided the best future for advertising. They concluded that agency development was the best way to achieve these objectives. Agency proponents asserted that the progression of the discipline would grow with the "expansion of the craft and agency professionalism" (Moriarty et al.).

Established agencies explored new ways to cultivate growth and development among clients and employees. For example, McCann created an agency philosophy that emphasized crafting strategic messaging, which was coined "truth well told" (Fox, 1997). Clients were reassured that advertising agencies undertook a systematic process that would yield fruitful profits. J. Walter Thompson (JWT) made strides in the realm of professionalism by creating the role of account executive. This occupational position served as a liaison between the client and the advertising agency to ensure an exceptional level of agency performance and credibility (Fox, 1997). Subsequently, JWT agency established a market research department and used psychology to inform advertising ("Ad Age Advertising Century: Timeline," 1999).

Other groups proposed the dominance of "atmosphere advertising," which featured opulent art and striking layouts. This approach appealed to newly emergent brands and class distinctions within society. Atmosphere advertising valued elegant writing and a complementary visual tone. A leading advocate of atmosphere advertising was Theodore F. MacManus, a prominent copywriter for General Motors. His famous advertisement The Penalty of Leadership was considered one the greatest advertisements of all time even 30 years after its debut (Fox, 1984). Like MacManus, creative practitioners often advocated atmosphere advertising as it increased their roles and the importance of imagination.

Atmosphere advertising was also supported by advertisers because it complemented emergent brands, class distinctions, and strategic targeting (Moriarty et al.). As advertisers desired to create mass audiences for mass consumption, strategic targeting was heavily advocated (Strasser, 2004). It was considered to be a great method to achieve financial efficiency and establish consumer connections. Advocates of atmosphere advertising believed it would increase sales through pairing the best consumer with the best product in an aesthetically appealing manner.

Despite a number of creative ideas and proposed directions, advertising did not reach a consensus. Dissension frequently resulted in gridlock. Seething discord was made readily apparent through futile unification attempts of trade organizations. Within the American Association of Advertising Agencies, major power blocs often disrupted unifying endeavors. For example, Albert Lasker, an executive at JWT who had become a powerful leader in advertising, refused to join any trade organization or association. In response to an offer, Lasker stated,

> We didn't see what we had to gain by joining the association. A lot of people joined the association that we didn't think had any right to be advertising agents, and we didn't want to seem to be, by being in membership, with them, giving approval to them ... Secondly we were getting a much higher price than anyone but two or three. We didn't feel it was our business to educate these others on how to get the higher price. (Fox, 1984, p. 69)

Acclaimed copywriter and advertising CEO Charles Austin Bates also revealed intense friction within the industry. Bates noted that within trade organizations, advertisers "came from newspaper offices, studios, the bar, and the pulpit; and they literally poured into the advertising arena a stream of delicious nonsense, which, if it could have been hardened, might have served for the decoration of afternoon tea cups" (Fox, 1984, p. 41).

Tumultuous Environment

As advertising coped with internal factions, there were numerous sociocultural factors that influenced the world around them. One major occurrence was the American temperance movement, a predecessor of the Prohibition era. Although alcoholic consumption was the core issue of the temperance movement, its scope soon broadened to include other social problems. Prohibition of alcohol was seen as a method of preventing crime, prostitution, and other abuses (Gusfield, 1986). These reform efforts were also considered a method of empowerment and became a vital component of additional movements that took place during the time (Fletcher, 2012).

Due to its range of influence, the temperance movement became intertwined within government policies, elections, territorial expansion, and economic models. As issues began to ferment, zealous reform efforts blazed through the nation. Consequently, the implementation of structures and systems that affected morality, power, and efficiency largely characterized society (Gusfield).

Although in a fledgling state, advertising was significantly affected by the influence of the reform area. Leaders worked vigorously to improve business practices and public perception of the field. They began with the creation and dissemination of fair and accurate media materials, including rate cards and information regarding circulation and outreach efforts. Contracts replaced handshakes and corrupt practices in order to ensure honest communication and interaction among advertising agents, advertisers, and media publishers (Fox, 1997).

Fervent reform in advertising coincided with the second wave of the Industrial Revolution. This massive technological shift began around 1850 and ended in 1914 when America entered World War I (Daunton, 1983). This period was characterized by the electronic transformation of industrial systems, mass production, and mass communications technology (Cawelti, 2002).

Progress achieved during this time was primarily based on scientific innovations and electronic inventions. America witnessed the birth of the transcontinental railroad, advancements in steel manufacturing, and the development of the assembly line. Mass production was made possible largely by machinery fueled by electricity. Moreover, hundreds of thousands of patents were issued for inventions based on electricity, including incandescent light bulbs, air brakes, and tabulating machines.

In concurrence with advances in electronification, advertising gained momentum. The success of mass-produced corporate products such as Kellogg's Corn Flakes, Ivory Soap, and Crisco shortening, helped advertising achieve wider public acceptance. In addition to benefiting from productivity, advertising advanced due to knowledge of personnel practices that accompanied electronification, namely the assembly line (Strasser).

Advertising leaders began to implement personnel systems, science, and knowledge informed by assembly line practices. N. W. Ayer established a "business-getting department" to plan advertising based on prospective marketing needs. Science was applied through the use of psychology ("Ad Age Advertising Century"). During this time, Dr. Walter Scott's *The Theory of Advertising* (1903) and *The Psychology of Advertising in Theory and Practice* (1908) were published. A central theme of Scott's works focused on personnel selection and practice.

America soon became an influential global leader in applied technology. Such advancement affirmed the importance of electricity in American society and abroad (Daunton). Achievement and productivity led to unprecedented growth in America. There were vast improvements in public health, sanitation, mass communication, transportation networks, and standards of living. Domestic and international trade grew as agricultural systems became fortified. Increased mechanization lead to more efficient methods to create products and generate revenue (Cawelti).

As new capabilities were harnessed, economic power and wealth began to grow. Major corporations, notably the American Telephone and Telegraph Company and the Standard Oil Company, began to rise in power and affluence (Marchand, 1985). Influential tycoons seized opportunities created by the tremendous growth of corporate America. Business leaders, including Andrew Carnegie, John D. Rockefeller, Cornelius Vanderbilt, Andrew Mellon, J. P. Morgan, Leland Stanford, and Charles Crocker, amassed legendary wealth and created powerful dynasties. America soared toward global dominance and developed a thriving middle and upper class (Cawelti).

Although America amassed tremendous riches, negative repercussions occurred as a result of rapid growth. Industrialization led to rapid urbanization and sharp declines in rural communities and the agricultural lifestyle. Conflicts erupted between rural and urban values. Labor structures, created to bolster new technological systems, decreased demands for unskilled laborers and agricultural workers. Unemployment and poverty rose as machines and factories displaced laborers. Fixed forms of capital became obsolete or endured shortened lifespans (Cawelti).

As society underwent significant shifts, urbanization issues continued to worsen. Cities felt the strain as more and more people relocated, including European and Asian immigrants, Americans seeking employment, and individuals attracted to urban lifestyle. These communities began to intermingle and coalesce. In turn, these diverse cultures produced unforeseen outcomes and several issues that demanded public attention. Critics argued that behaviors and lifestyles that accompanied industrial areas were undesirable.

Cities were associated with crime, morally corrupt behavior, prostitution, alcoholism, and other abuses. Urban critics used politics and social-reform efforts to elicit the change they desired (Cawelti).

During this time, the rights of the disadvantaged in cities were championed by influential organizations. The women's suffrage movement voiced concerns regarding prostitution and other female offenses in cities. Its leaders advocated equality, empowerment, and the right to vote for women. Religious missionaries heavily supported social activism crusades with the Salvation Army and YMCA. Labor unions sparked intense battles as they worked to ensure fair wages and safeguards against increased competition and unethical business practices (Fletcher).

Like urban areas, Southern communities experienced turmoil as well, including economic setbacks and racial violence. These issues were further compounded by the transition from agrarian to industrialized communities. Consequently, intense social and civil unrest ensued. Despite promises of freedom and lifestyle improvements, African Americans continued to endure terrorism and inferior treatment in the South. Jim Crow legislation effectively infused structural racism and violent oppression into every facet of society for African Americans (Inwood, 2011). Advertising reinforced deplorable conditions involving this community. Content and imagery depicting racism, subservience, and heinous abuse were frequently used in advertisements (Jenkins, 2014).

Reform movements created improvements for some individuals and exacerbated conditions for others. For example, many urban activities that improved the lives of the wealthier citizenry directly impacted sources of capital and increased power for the newly emergent wealthy class. Yet, while wealth increased for certain segments of the population, creating a class of affluent business owners, these same individuals argued with social reform advocates. Therefore, sharp divisions and intense competition began to characterize politics of the period. Leaders of opposing sides viewed major governing bodies, including the presidency, congress, and Supreme Court, as ways to exert their influence and control (Gusfield).

Even as sharp divisions between wealth and poverty in America became an issue of major concern, many rich and powerful were accused of unethical, exploitive, and monopolistic practices that resulted in impoverished conditions and stratification of wealth. As gaps persisted, the nation's responsibility to the poor and disadvantaged became a topic of intense debate. A new agenda began to emerge that intensified reform efforts (Gusfield).

Leaders worked to make sense of the new world that emerged after the Industrial Revolution. As America conceptualized frameworks for corporate policy, social responsibility, the economy, and the rights of the poor, varied ideologies swirled within the public sphere (Daunton). The intermingling of discoveries in physical science and social physics were highly influential. The coalescence of these ideas characterized the economic, political, and social worldview of the time (Carey, 1960).

This synthesis of knowledge is readily seen through the works of Sir Isaac Newton. Although published in 1687, Newton's *Philosophiæ Naturalis Principia Mathematica* (*Mathematical Principles of Natural Philosophy*) remained dominant and highly influential for more than three centuries. This work provided the foundation for the Age of Enlightenment, as well as the Industrial Revolution. Newton's three universal Laws of Motion enabled advances that led to new and efficient methods for power sources, mechanization, and mass production (Carey, 1960).

Referred to as the Newtonian World Machine, Newton offered a collection of knowledge that explained mechanistic systems, harmonious order of the universe, natural laws governing phenomena, atomistic conception of matter, as well as the ability to reason and the comprehension of reality. These discoveries in physical science provided a cornerstone for the works of numerous influential philosophers, including John Locke (Carey, 1960).

In the same way that Newton conceptualized the atom as the basic particle of the harmonious physical world, Locke deemed that man was the basic unit of society. Like atoms, Locke believed man had the innate ability to develop a society in accordance with natural order and create systems to exercise reason (Carey, 1960).

Locke's philosophies were a central component within egalitarianism and other moral doctrines. Akin to atoms in Newton's system, man was considered equal by nature. Previously, society was dictated by a bound system of fixed status relationships. In a revolutionary departure from widely held beliefs, intellectuals suggested that environmental factors and opportunities for cultivation characterized the development of human beings rather than tradition-bound systems (Feingold, 2004).

These beliefs gained momentum in the Age of Enlightenment (approximately 1700–1800). The movement emphasized the use of intellect and reason to reform society. Ideologies grounded in religion, intolerance, superstition, and abuses by the church and government were vehemently challenged.

Scientific method, skepticism, and intellectual interchange were promoted during the Age of Enlightenment. This movement began in Europe and later spread to America. The ideas of the Age of Enlightenment had major impact on culture, politics, the economy, and government (Feingold).

The Age of Enlightenment popularized Locke's doctrine of natural rights, which concluded that man originally lived in a state of nature and possessed the natural rights of life, liberty, and property. It was proclaimed that these rights are sacred and inalienable for all men. Within a society, man creates institutions, systems, and governments to ensure communal harmony, but does not and cannot surrender his natural rights (Carey, 1960). The natural right of property indicates that any man is entitled to any object in which he contributes his labor. There is an inseparable connection between individual freedom, rights, and property. Partial ownership of land is established through a man's labor (Carey, 1960).

During the Industrial Revolution, intellectuals challenged that these same concepts and rights would translate to corporations, urban cities, and complex mechanized work systems. Many of these issues fueled moral concerns, labor disputes, and reformation agendas. Simultaneously, further advancements in scientific knowledge coalesced once more with philosophical orientations (Feingold).

In 1859 Charles Darwin, a natural historian and scientist, published his theory of evolution in his controversial book, *On the Origin of Species*. Darwin asserted that all species of life descended over time from a common ancestry. Darwin's work popularized the concept of natural selection, a process by which species with desirable traits are systematically favored for reproduction (Claeys, 2000).

In essence, Darwin's concepts held that a natural environment of a species "selects" traits that affirm reproductive advantage. This theory was in stark contrast to a belief in creationism as perpetuated by Christianity. Like Newton, the works of Darwin became enormously popular and highly influential. They were applied to various fields of study, including psychology, economics, culture, and politics (Claeys).

Philosopher Herbert Spencer expanded on concepts of natural selection attributed to Charles Darwin. Spencer used Darwin's ideas to create a foundation for explaining the stratification of wealth and poverty as a natural occurrence in society. Spencer coined the term, "survival of the fittest," in *Principles of Biology*, published in 1864. Spencer's works were highly influential in economics and policy. Interpretations of his concepts revealed a belief that population segments would naturally adapt to their environment for survival. Conversely, weaker population segments would not be able to adapt. Eventually, society would be comprised of elite population segments and, therefore, function on high levels.

Proponents of such ideology deemed it unnecessary to help individuals who struggled within society. Arguments were crafted to create social policies that made little distinction between those able to support themselves and those who were not able to do so. Thus, reform movements to assist urban populations with economic, educational, and social inequities began to face major opposition.

These philosophies became evident within other vital aspects of American society. Capitalism was subsequently adopted under the premise that it would be governed by the same harmonious mechanistic order as the physical and social world. By the same token, laissez-faire economics ensured that the market was free to self-regulate through competition. Intellectual development ensured that neither the church nor government would impose formidable restrictions.

Self-regulating mechanisms of the economy were bolstered through ensuring that all entrants into the market would have adequate knowledge of regulatory practices. Each individual was intended to have information regarding pricing resulting from supply-and-demand relationships. The responsibility to supply such information was assumed by the market (Carey, 1960). Accordingly, providing this information to consumers was the premiere function of advertising. However, as the environment swirled with numerous issues, needs, and concerns, advertising began to assume additional responsibilities.

As advertising developed, its core and functions were shaped by the social, political, economic, and intellectual climate of the time. A range of events, controversies, and dominant ideologies, which included social reform, the Industrial Revolution, urbanization, and laissez-faire economics, characterized the conception of advertising. The impacts of such influencers became evident in subsequent practices and central operations.

In addition to external factors that influenced advertising, there were a number of internal shifts that shaped its development as well. Overall, the enveloping themes of morality, power, and productivity remained ever present. The implementation of ethical behavior, contracts, and standardized procedures reflected moral reform efforts. Advantageously, public perception of advertising began to improve. The creation of advertising agencies affirmed the power and the spirit of progressive entrepreneurism that characterized the times. With systematic operations, improved public perception, and increased power advertising began to make progress. Yet, there was still no consensus.

· 2 ·

DRAMATIC DIFFERENCES

Advertising functioned as a fractured and disconnected community for more than 50 years after its inception. Although significant progress had been achieved, advertising agencies failed to gain the traction needed in order to move forward collectively. As a result, advertising struggled to become a full-fledged profession. Two main factors inhibited collective progression in advertising. The first factor was that there was no established paradigm to unite advertising. Second, power was highly concentrated among close-knit groups.

The absence of an established paradigm in advertising could be attributed to a lack of widespread knowledge. Advertising itself was relatively young. Consequently, there were diverse approaches to advertising and virtually all of them appeared valid. There was no standard upon which to determine the credibility or validity of varied approaches to advertising. Without a unifying paradigm there was an inability to distinguish facts from opinions. In addition, the knowledge and technology relied upon by advertising was in flux. Accordingly, advertising was fragmented and unstable. It had yet to ascend to a level of sophistication that could sustain a universal enterprise.

Power within advertising assumed several forms, including knowledge, wealth, and relationships. The concentration of these elements prevented a diffusion of power throughout the advertising community and inhibited collective progression. For example, lucrative conglomerates created their own in-house advertising departments or conducted business exclusively with major agencies. This often resulted in a concentration of wealth among these parties. Largely due to affluence, these parties were equipped to invest in research, technology, and systematic operations. In turn, the most talented and innovative advertising professionals were often attracted to and recruited by conglomerates and major agencies.

An example of the concentration of power in advertising is revealed through the history of the New York Telephone Company. This business endeavor was the result of a lucrative arrangement between inventor Alexander Graham Bell and his financers, Gardiner Hubbard and Thomas Sanders. The New York Telephone Company achieved success with the aid of advertising. They created one of the largest in-house advertising departments of the time and used sophisticated advertising methods to effectively target consumers. The New York Telephone Company also employed Harry McCann, a savvy and influential advertising executive. McCann led experimental advertising campaigns that allowed it to connect with mass consumers (Marchand, 1985).

McCann's success at the New York Telephone Company attracted the attention of the Standard Oil Company, one of largest and most lucrative companies in the world. John D. Rockefeller, financier and American business magnate, was a founder and premiere shareholder for the Standard Oil Company. In 1911 McCann was hired to lead the in-house advertising department of the Standard Oil Company. However, in 1912 the Supreme Court ordered the dissolution of John D. Rockefeller's Standard Oil empire. McCann then formed his own self-titled advertising agency, which would grow to become a highly influential global success.

Like other major advertising leaders of his time, McCann was privy to wealth, knowledge, and clients as a result of powerful relationships. Many of McCann's most lucrative clients were the subsidiary oil companies that resulted from the dissolved Rockefeller empire (Johnson, 2011). Moreover, McCann's experiences and prowess led him to implement some the most sophisticated and successful methods used in advertising during his time. McCann introduced the total marketing concept in 1912, which led to overseas growth. He also implemented the use of public relations, research libraries, sales promotion, and production survey services in advertising.

In addition to McCann, several major advertising agencies, including JWT and N. Y. Ayer, achieved success through solid relationships with noteworthy clients (Johnson, 2011). For more than half a century a concentration of knowledge and wealth existed that was not disseminated widely throughout advertising. Close-knit relationships contributed to power blocs and perpetual cycles of centralized affluence and productivity.

A Turning Point

In 1917 advertising reached a turning point. The American government officially recognized advertising by establishing the Advertising Division within the Committee on Public Information (CPI). Through executive order, every branch of advertising was asked to cooperate. The Advertising Division was instructed to bring various elements of advertising together and discover methods to harness the best advantages. Members of the advertising community were deeply gratified and delighted to be included in government endeavors. They worked diligently and exceeded all expectations. This marked a turning point for advertising, after which things would be dramatically different.

The year 1917 also marked America's entrance into World War I. This was the first national total war experience for both America and the advertising community. President Woodrow Wilson commissioned the George Creel Committee on Public Information (CPI) to disseminate information pertinent to the war. Referred to as "the world's greatest adventure in advertising" this independent agency's objective was to shape and influence public opinion in support of American participation in WWI (Fox, 1997, p. 75).

The CPI used virtually all forms of mass media, including newspapers, radio, telegraph, cable, and film. It also created posters, novels, guerilla tactics, and rhetoric to persuade public opinion. Moreover, the CPI contained more than 20 bureaus and associations, notably the advertising division (Creel, 1920).

According to *Printers' Ink*, an advertising trade journal, leaders of the Advertising Division were prominent figures in advertising and trade organizations. W. H. Johns was president of the American Association for Advertising Agencies, and vice president of George Batten Advertising Agency of New York. William C. D'Arcy was president of the Associated Advertising Clubs of the World and president of the D'Arcy Advertising Agency of St. Louis. L. B. Jones was president of the Association of National Advertisers and advertising manager of Eastman Kodak Company in Rochester, New York. Herbert S. Houston was chairman of the National Advertising Advisory Board of the Associated Advertising Clubs of the World and vice president of Doubleday

Page & Company of New York. O. C. Harn was chairman of the National Commission of Associated Advertising Clubs of the World and advertising manager of the National Lead Company in New York.

Members of the division of advertising also represented a range of diverse influence and expertise. As specified through their official commission they sought cooperation from every branch of advertising. Accordingly, members represented media, publishers, agriculture, newspapers, business, international, and national advertising. Additionally, subcommittees were created to ensure cooperation between members and the leadership of the Advertising Division. Subcommittees were also diverse and included a range of interests and expertise, including posters and periodicals.

Under the guidance of prominent members of the field, the Advertising Division placed more than $1.5 million of donated ad space and copy (Fox, 1997). Additionally, private advertisers featured war themes and appeals to civil liberty in their campaigns. The use of dynamic visual imagery, graphic typography, memorable headlines, and informative copy created iconic ad campaigns. Successful messaging was created that appealed to mass audiences, as well as specific niche communities (Creel, 1920).

Although controversy began to surround the CPI involving the suppression of civil liberties and reform movements, advertising made tremendous strides as a result of its participation. Knowledge was diffused through cooperative efforts. A consensus was reached regarding the most effective advertising methods. Additionally, the perception and image of advertising soared to new heights.

Following the formal dismantling of the committee in 1919, George Creel professed the following regarding the Creel Committee's efforts during the war:

> The work, as a whole, was nothing more than an advertising campaign. And I freely admit that success was won by close imitation of American advertising methods and through the generous and inspirational cooperation of the advertising profession. Before the war, your status was anomalous. Today, by virtue of government recognition as a vital force in American life, you stand recognized as a profession. (Fox, 1984, p. 76)

The New Republic declared that the advertising man was "the cornerstone of the most respectable American institution (and that) the newspapers and magazines depend on him; Literature and Journalism are his hand maidens. He is the Fifth Estate" (Fox, 1984, p. 77). *Printers' Ink* affirmed, "Advertising has earned its credentials as an important implement of the war" (Fox, 1984, p. 75). A few years prior, advertising agents were considered untrustworthy and a tolerated nuisance. In an interview with *Advertising Age* in 1930, Daniel Lord, a leading advertising practitioner affirmed the dramatic contrast. He recalled that he and many of his peers once owed their lives to legislation that prohibited killing advertisers (Fox, 1997).

Public support was further strengthened by wartime fervor. National advocacy quelled many reform movements of that time. Accordingly, many of advertising's harshest critics were eliminated. Advertising became associated with actions toward public good. Improved public perception helped catapult advertising toward prosperity and growth. World War I served as an impetus for significant change. As a direct result of the knowledge, performance, and success attained during World War I, advertising emerged as a full-fledged profession.

A Unifying Paradigm

Acquiring a unifying paradigm was one of the most profound outgrowths advertising has experienced. It provided the foundation for future growth and expansion. While solving one of the most pressing issues faced by its community, the most effective methods of advertising were revealed. At the same time, collaborative efforts and cooperation helped diminish factions and rivalries, which cultivated an opportunity for consensus.

During World War I, members of the advertising community gravitated toward one conceptual framework. Although there were a plethora of ideas from which to choose, the emanatory had to be consistently and effectively successful. Advertising's executives, practitioners, and researchers created impressive campaigns to sell war bonds, raise funds, boost morale, enlist military recruits, disseminate information, promote conservation of food and resources, and other activities (Tungate, 2007). The United States produced more propaganda posters and materials than any other nation in the war. A spokesperson for the J. Walter Thompson advertising agency revealed that the war gave advertising men an opportunity to render valuable patriotic services and to reveal to a wide circle of men the real character of advertising and the important function it performs (Marchand, 1985, p. 6).

The unifying paradigm blurred "the reason why" tradition with cohesive campaign techniques. Prior to World War I Albert Lasker of JWT proposed the reason why as a leading approach. It was known for its ability to drive consumer behavior. Later, as a result of various agencies participating, another distinct mode of advertising expression developed that was unlike anything before.

The institutional paradigm emerged and was established as the basis for advertising (Hovland & Wolburg, 2010). This approach was anchored in exerting influence rather than simply providing information. It relied upon the power of persuasion. Under the new paradigm two distinct methods were primarily used to accomplish this goal. The first method involved collaboration through interdisciplinary efforts. The second method applied the strategic use of modernity. Last, a firm appreciation and respect were acquired for the power of visual imagery and emergent technology. These elements became the cornerstone of advertising tactics. Moving forward, advertising would build upon visual communication and the ability to harness technology for much of its success. Upon this dynamic foundation, advertising transitioned to a period of prosperity.

Interdisciplinary Collaboration

Successful collaborations during wartime efforts led to increased interdisciplinary practices. Prior to this time, advertising was often separated into distinct branches. Advertising functioned within silos and projected ideas from a narrow scope. Forced collaboration created an exchange of ideologies and practices and the diffusion of knowledge led to sophisticated techniques. Research, business, and graphic communication emerged as key areas of collaboration. The acquisition of harmony among these elements continually resulted in advertising success. Although the elements and degrees of hierarchy within collaborations varied according to specific objectives, it became clear that interdisciplinary collaboration was critical to advertising.

Under the key area of research, science came forth as a useful tool for advertising. Research and science were used to develop strategies and influence behavior. It became increasingly known that social sciences, including psychology and sociology, were extremely useful when applied to advertising. *Printers' Ink* affirmed that wartime advertising demonstrated that "it was possible to sway the minds of whole populations, change their habits of life, create belief, practically universal, in any policy or idea" (Marchand, p. 6).

Psychology was used to facilitate advertising's ability to differentiate products. This was achieved primarily through the use of slogans, positioning, prestige, and association (Ohmann, 1996). Prior to interdisciplinary collaboration, differentiation was based on physical attributes such as product quality, features, and price. Based on its new paradigm, advertising began an enduring practice that maximized appeal and minimized information (Potter, 1960).

Successful interdisciplinary collaboration led to the postwar employment of leading psychologists within advertising agencies. For example, John B. Watson was hired by JWT, one of the most influential agencies in advertising. Watson, a founder of behaviorism psychology, asserted that he had discovered the basic techniques for predicting and manipulating human actions (Goldstein, 2011). He promised advertising colleagues that they could make consumers react. In order to do so, Watson maintained that it was "necessary to confront consumers with either fundamental or conditional emotional stimuli" (Fox, 1984, p. 85). While at JWT, Watson conducted research that refuted informational approaches as a basis for advertising. He gained additional notoriety by advocating behaviorism through demonstrations and rhetoric. Watson became an account executive and was promoted to a vice presidency at the JWT agency (Fox, 1984).

The prominence of JWT, and its acclaimed agency roster, made the inclusion of psychology within advertising desirable. Accordingly, other agencies began to emulate similar practices. Advertising, informed by behaviorism, relied upon the exploitation of human drivers and passion points, including love, vanity, jealousy, fear, and rage.

In addition to prestigious psychologists, leading agencies employed the use of academics and social psychologists as well. Advertising practitioners noted that in the spirit of emulation, they wanted to replicate the works of those deemed superior in taste, knowledge, or experience (Fox, 1997).

The theories of Dr. Walter Dill Scott, of Northwestern University, were applied. In his book, *The Psychology of Advertising*, Scott asserts that customers are typically convinced to purchase due to suggestions directed at the proper psychological moment (Fox, 1997). Advertising practitioners were eager to apply these social science models. Influential advertising executive B. L. Dunn proclaimed that nearly every important decision made by humans is made within the subconscious. Practitioners also implemented the theories of Sigmund Freud and Carl Jung. These theories were applied to create criterion to inform advertising campaigns, marking an intrinsic partnership that would be evident for years to come (Goldstein).

Yale graduate and scholar Stanley Resor used his academic knowledge and business expertise to help structure agency life. Resor became well known for his formidable contributions to JWT. He implemented the use of strategic planning within campaigns and spearheaded much of the formidable work involving target audiences. Resor helped establish two enduring aspects of advertising: the coining of brand names and the education of a given social class used to facilitate the imitation of the habits of the rich and elite (Ohmann).

The work of esteemed social scientist and psychologist Cesare Lombroso helped structure gender roles in early advertising. Lombroso claimed that women possessed a stronger ability to excite their imagination with external objects (Fox, 1997); therefore, the desire to emulate was deemed stronger in women than in men. Thus, products and campaigns during that time were largely targeted toward women. This tradition became a cornerstone of modern advertising.

Due to its success and future promise, interdisciplinary collaboration became essential to advertising. Interdisciplinary collaboration created methods to sell the abundance of industrial products. Interdisciplinary collaboration also established markets and cultivated ideal consumer behavior. By 1900 advertising's budget was steadily climbing. By 1918 it reached $1.5 billion, which was a tremendous increase when compared to years prior. Anchored within the institutional paradigm, the future would soon reveal that advertising would soar to even greater heights. *Printers' Ink* proclaimed:

> This is a golden age in trademarks—a time when almost any maker of a worthy product can lay down the lines of a demand that will not only grow with years beyond anything that has ever been known before, but will become, in some degree a monopoly ... Everywhere ... there are opportunities to take the lead in advertising—to replace dozens of mongrel, unknown, unacknowledged makes of a fabric, a dress essential, a food, with a standard trade-marked brand, backed by the national advertising that in itself has come to be a guarantee of worth with the public. (Potter, p. 21)

Strategic Use of Modernity

In addition to interdisciplinary collaboration, advertising success was also attributed to the strategic use of modernity (Marchand). Although crude by contemporary standards, advertising practitioners conducted research to determine the best methods to meet the demands of the mass market. A transformation of culture had to take place in which ideas, lifestyles, relationships, and practices at the local level shifted toward mass production and national markets.

This conversion began prior to World War I; however, the strategic use of modernity during the war jolted this transition into practice (Strasser, 2004). Advertising applied the strategic use of modernity in two main ways, through content and visual imagery.

World War I demanded propagandistic materials and practices of irrefutable clarity and didactic messaging. Consequently, formidable social and cultural norms were eradicated. Taboo topics, unleashed through advertising, created paradoxical conflicts in numerous areas, namely class deregulation, gender roles, male virility, sexual behavior, and morality (Lathan, 2006). The vast geographic sweep of World War I created enormous change and a heightening of modern systems (Mackaman & Mays, 2000).

In addition to subject matter, modernity was bolstered through the use of visual imagery. The abundance of mass production and consumption became a celebrated aspect of war (Mackaman & Mays). Boasting of the size of output and distribution became a frequent occurrence. Copy voice and tone also insinuated that the modern public purchased mass products, which created audience desire. Depictions of factories, production equipment, illustrations of buildings, machines, and workers were used strategically as well.

Additionally, advertising often disproportionately emphasized styles, behavior, practices, and technologies that were new and ever changing, thus, foreshadowing a function that would enhance the fidelity of their craft (Strasser, 2004). Postwar, the advertising industry sought to sustain progress. Previously, strategies ranged from the use of magic through the depiction of elves and fairies to the use of wisdom through the depiction of advice from the elderly (Ohmann). Modernity eventually advanced to the forefront as the most successful and consistent strategy.

Modern Advertising (1905), published by Calkins and Holden, provided countless examples from manufacturers, advertising practitioners, salesmen, and publishers that attested to the success of this strategy. It employed the use of surveys, statistics, and interviews to convey motivational trends of consumer brand selection, levels of attention, buying habits, and so forth. *Modern Advertising* also offered advice on best practices and criticism of advertising to spur improvements. Moreover, later editions of the book were informed by the psychological theories of Scott and Gale and the effects of wartime fervor (Marchand). These updated manuscripts were rereleased and frequently utilized by advertising practitioners (Moriarty, Mitchell, & Wells, 2012).

The strategic use of modernity became critical to the future of advertising. Although the American population skyrocketed, industrial systems produced as much as fourteen times the amount of raw materials needed to manufacture goods (Ohmann). Moreover, newly emergent factories were developed based on the model of flow production. This process enabled enormous amounts of raw materials to be processed automatically in a constant stream (Potter).

America began a new age as an industrial nation, signaling the shift from a primarily agrarian to an industrialized economy. Many of the outcomes of this situation helped to strengthen advertising as well. The rise of mass production, modernized transportation, broad distribution, branding, and urbanization all helped to strengthen advertising. Additionally, fewer Americans possessed land and resources to produce goods. Americans began to purchase manufactured products at soaring rates (Hovland & Wolburg).

Through the use of demand creation, advertisers sought to increase desire among consumers in order to meet and exceed expectations for purchase within mass markets. Accordingly, the strategic use of modernity spurred the concept of planned obsolescence (Hovland & Wolburg). American advertising began to nurture the idea that whatever service or product could be purchased yesterday is far inferior to anything that could be purchased today. Advertising practitioner and publisher of *Modern Advertising* Elmo Calkins asserted:

> The purpose is to make the customer discontented with his old type of fountain pen, kitchen utensil, bathroom, or motorcar because it is old fashioned, out-of-date. The technical term for the idea is obsoletism. We no longer have to wait for things to wear out. We displaced them with others that are not more effective but more attractive. (Hovland & Wolburg, p. 31)

Obsoletism was intended to be beneficial to both consumers and the overall economy (Strasser, 1994). Advertising practitioners initiated the application of modernity and its spurred concepts. However, these ideas began to expand and greatly contribute to America's perceived abundance. Continuous consumption, facilitated by the advertising industry, became instrumental in sustaining the American capitalist economy (Potter). In order for a capitalist society to fully operate and prosper, there must be continuous production and consumption. Hence, the advertising industry and its practitioners further solidified their roles, functions, and importance in America and the impending global spectrum. Through selling ideas and stimulating the demand for goods, advertising practitioners revealed themselves as geniuses behind a newly discovered economic perpetual-motion machine (Marchand).

Advertising practitioners proudly proclaimed themselves missionaries of modernity (Marchand). Unabashedly, practitioners created a bias within the American public for the new against the old and the modern against the old-fashioned. They often played a therapeutic role as the American public adjusted to seemingly incessant change. As problem solvers, they strove to find methods to engage consumers by making them feel comfortable with and ultimately purchase products (Fox, 1997). In essence, advertisers gave increasing attention to the consumer as opposed to the product. Practitioners persuaded consumers to live vicariously through experiences with products by detailing vignettes of social life. Marchand affirms the significance of these newly crowned apostles of modernity:

> Inventions and technological applications made a dynamic impact only when a great mass of people learned of their benefits, integrated them into their lives, and came to lust for more new products. Modern technologies needed their heralds, advertising men contended. Modern styles and ways of life needed their missionaries. Advertising men were modernity's town criers. They brought good news about progress. (Marchand, p. 1)

In the year following the success of the George Creel committee, annual advertising volume doubled, reaching nearly $3 billion. Companies invested exponentially in advertising and acquired full-page layouts in newspapers, double-page inserts in magazines, and elaborate illustrations. In 1919, an advertising executive asked the discipline that, in forging ahead, "are we going to rest upon the record of advertising as a factor of the war or are we going to develop it still further, to apply it to the many fields in which it can serve?" (Fox, 1984, p. 78).

· 3 ·

GROWTH & EXPANSION

In the 1920s advertising entered its second phase of development. This period was marked by exponential growth and expansion. Advertising built upon the unprecedented success it achieved through its participation with the Committee on Public Information (CPI). The knowledge they acquired provided a paradigm for future prosperity.

Although pivotal, the newly established paradigm was limited in scope and precision due to its infinitesimal state. Consequently, advertising devoted significant attention to its articulation. Advertising's primary objective was to establish methods to increase productivity and effective problem solving using the new paradigm. Advertising also sought techniques to reinforce new understandings. This was accomplished in several ways. The first approach involved discovering methods to apply the paradigm to produce consistent successful outcomes. Advertising also explored ways to produce data and systems of measurement. Moreover, advertising investigated ways to identify techniques to apply the paradigm to future areas of interest (Kuhn, 1964).

As a full-fledged profession advertising was challenged to satisfy the requirements of an entire discipline. This included a number of areas. The various demands of advertising, including business, research, and graphic communications as well as the needs of advertising professionals, including executives, practitioners, scientists, and advertisers, had to be met. Trade journals and scholarly publications had to be sophisticated and were promoted throughout the discipline. Additionally, curriculum had to be established to cultivate a thriving academe.

Although advertising was tasked with a number of goals, the profession was invigorated by confidence and favorable conditions. As a formidable contributor to the near unanimity of public support for WWI, advertising was entrusted with a new and bold sense of power (Fox, 1997). *Printers' Ink* affirmed "advertising had earned its credentials as an important implement of war" (Fox, 1984, p. 75). At the onset of the 1920s, advertising was poised and prepared to fulfill its promise. It passionately embarked on the work needed to affirm the newly established paradigm.

Cultural Fusion

The 1920s serve as the benchmark for the inextricable relationship between pop culture and advertising. This alliance resulted in cultural fusion. Through advertising American citizens learned to become consumers. A raw population accustomed to homemade goods, bartering, and unbranded merchandise was converted to a national market (Strasser, 2004).

In tandem, advertising altered society. During the Roaring Twenties, culture and communications became increasingly organized around marketing consumer goods, mass production, and consumption (Strasser, 2004). "Amid these fluid circumstances, advertising took on new powers … It changed American life down to its most intimate details, with a speed and totality that left observers groping for precedents" (Fox, 1984, pp. 78–79).

Much of this fervor was fueled by advertising's desire to affirm its new profession. Armed with knowledge, success, and credentials, advertising set out to articulate its paradigm. Accordingly, advertising during the 1920s was largely characterized by prosperity and cultural fusion. Prosperity affirmed the successful articulation and application of knowledge, while cultural fusion foreshadowed entanglements associated with inextricable complexity. As advertising rose in cultural prominence, there were also shifts in societal values and the perception of consumers.

Success achieved throughout the 1920s affirmed the new paradigm. Both scholars and practitioners created works to further its development. This resulted in the strengthening of the profession. Useful theories, methods, terminology, data sets, ideologies, and campaigns were created. Advertising flourished, which led to increased productivity and insights. The institutional approach to advertising garnered sizable budgets and a reverence for the sensitivities of human desire.

In 1923 acclaimed copywriter Theodore MacManus helped Walter Chrysler achieve success with the use of advertising campaign techniques (Curcio, 2000). This groundbreaking achievement launched Chrysler Corporation's first car, the Chrysler. The campaign included a unified theme, a coordinated rollout, strategic media placement, and teasers. MacManus incorporated advertisements that featured interviews with Walter Chrysler. This strategy is a testimonial approach in which the company's owner or a staff member acts as a spokesperson for the product. This approach is used to increase trustworthiness and authentic connections with consumers. Although MacManus used a hard sell to identify the product features, the overarching theme of the campaign was driven by prestige and romance (Curcio).

Surges in productivity propelled advertising into the decade yielding its greatest influence on American life that resonates even today (Hovland & Wolburg, 2010). Wealth, accrued from products created during these times, became an economical driving force nationally and internationally. In this way the Roaring Twenties signified advertising's coming of age.

As advertising began to establish itself as a cultural icon, its relationship to the operation of America's free market system became increasingly complex. Upon inception, advertising had the function of supplying market information. However, as advertising developed it began to apply more persuasive techniques. There was concern that conflicts would arise that would result in biases that would negatively impact consumers. Potter (1960) affirms the transformed functionality as follows:

> Producers were no longer trying merely to use advertising as a coupling device between existing market demand and their own supply; rather, they were trying to create demand. Since the function of advertising had become one of exerting influence, rather than one of providing information, the older factual, prosy notice, which focused upon the specifications of the commodity, now gave way to a more lyrical type of appeal, which focused instead upon the desires of the consumer. (p. 22)

As advertising increased in importance, it also expanded its realm of influence. Through working to affirm the newly accepted paradigm, advertising took strides toward dominating media, harnessing technology, and influencing societal norms. Its ability to shape and influence popular standards placed advertising on par with America's most prominent institutions. Hence, the ideas, values, and habits surrounding advertising increased in significance.

As perceptions of advertising shifted and grew in importance, there were shifts in how consumers were perceived as well. Previously, consumers were generally perceived as rational human beings within the field of advertising. Consequently, advertising was primarily informational and reinforced the belief that consumers would make the most sagacious choice (Ohmann, 1996). However, as advertising began to function under its new paradigm it employed techniques associated with behaviorism and self-gratification. Ads became increasingly persuasive and appealed to the desires of consumers (Hovland & Wolburg).

Technology & Turmoil

Advertising productivity continued into the 1930s. With new knowledge, techniques, and experience with technology, advertising had the ability to turn previous failures into success.

When radio technology was first introduced it was a noncommercial medium and heavily rebuked by advertising (Socolow, 2004). Advertisers were reluctant to air content during the initial stages of radio. This presented a perplexing conundrum. Radio placed a renewed emphasis on words without visual images. Additionally, radio demanded the incorporation of sound effects, music, multiple voices, and ambient clutter in order to be memorable (Fox, 1997). However, under the accepted paradigm, radio was eventually embraced by the advertising community and grew into significant importance.

Advertisers recognized that radio supplied an invaluable opportunity to seize an additional segment that would strengthen its mass audience. Prior to the use of radio in this capacity, advertisers relied primarily upon print media to reach a mass audience. Newspapers generally reached local audiences (Strasser, 2004). Movies appealed to urban working-class communities (Lathan, 2006). Tabloids and magazines enthralled the sensibilities of what was deemed lowbrow popular culture (Taylor, 2009). Radio facilitated the opportunity to capture wealthier audiences of cultural significance. Upon inception, radio technology ownership was primarily concentrated among an elite audience and gradually spread to less affluent classes. Moreover, radio sets were predominantly located within homes, which represented an untapped market. Radio could also amass preliterate and illiterate audiences (Danesi, 2012).

Radio technology allowed advertising to increase and expand not only its depth but also its range of influence (Hovland & Wolburg). In accordance with the newly established paradigm, the capabilities of radio were an invaluable asset. Advertising practitioners worked diligently to harness radio's attributes (Marchand, 1985). Program sponsorship, product testimonials, and endorsements were utilized as successful techniques. By the close of the decade, the lines between programming and content blurred significantly. Agencies developed in-house radio advertising departments and previous government restrictions virtually disappeared (Hovland & Wolburg).

Advertisers celebrated the interdependencies of modern society. They sought to further rationalize the operations of the marketplace, lubricate its mechanisms, and achieve greater control over advertising's newly emerged functioning (Potter). The advertising industry's use of radio helped to further articulate the promise of the discipline. It extended the knowledge and application of the paradigm by correlating predictions with behavior and events. Although American industrial society had matured, the consumer remained the most unpredictable and the most disruptive element in the economic system. Advertisers worked to induce consumers to comply with the needs of the market through a dependency on products. Moreover, advertisers sought to educate and condition consumers toward a predictable and enthusiastic demand for new products, thus, enhancing the rationality and dynamism of the modern business system and the advertising discipline itself (Marchand).

The alliance between radio and advertising proved to be even more beneficial as time progressed. Radio gave the advertising industry a much-needed boost during harsh economic times (Fox, 1997). The stock market crash of 1929 signaled the beginning of what became known as the Great Depression, a catastrophic worldwide economic downturn. America was engulfed in severe turmoil. While the 1920s were exceedingly prosperous, the early 1930s were treacherous worldwide. Unemployment was rampant, stock prices plummeted, trade suffered, and famine grew (Hovland & Wolburg).

Although the exact cause of events that spurred the Great Depression remains unknown, the public blamed advertising (Marchand). It was claimed that advertisers seduced the consumers with lavish excesses. The industry counteracted with no frills advertising and campaign themes to mitigate and assuage anxieties. Nonetheless, many agencies slowed production, cut salaries, and eliminated staff. The advertising industry then turned much of its efforts toward research and the further articulation of its discipline. During this period, Daniel Starch, A. C. Nielsen, and George Gallup founded companies devoted to advertising research (Fox, 1997).

War & Peace

In 1939 World War II began, signaling the beginning of the end of the Great Depression. As with World War I, the advertising industry was bolstered from American participation in war (Young, 2005). In 1942 the War Advertising Council was founded. However, unlike during World War I, advertisers were no longer as eager to devote their time, energy, and enthusiasm. Advertising was now an established discipline with well-respected professional efforts.

Esteemed advertising practitioner Raymond Rubicam suggested that the recruiting of men and money for wartime efforts was not undertaken in an atmosphere of universal agreement. There was some dissention and criticism. Advertising executive Bruce Barton noted that advertisers simply set forth in pictures and copy the administration's argument, which was sound, patriotic, and moral but did not tell the truth. Nonetheless, the industry donated nearly $1 billion in space and time to World War II efforts. Advertising expenditures grew from $2.2 billion in 1941 to $2.9 billion in 1945, the end of the war (Fox, 1997).

Much like the postwar period after World War I, advertising grew as a result of its wartime involvement. Advertising specialists attained new knowledge and technological experiences (Young). One of the most significant achievements applied in advertising came in the form of a mass communications model. The Shannon-Weaver model of advertising (1948) was a key achievement that was used by the military for cryptanalysis in World War II (Khan & Baig, 2007). This model proposed that information was communicated by sending a signal through a sequence of linear stages (Shannon, 1948).

The Shannon-Weaver model became widely accepted and provided the foundation, as well as a unifying general theory, for mass communications. Highly influential, it stimulated subsequent works in modeling, education, research, and practice (Pietila, 2005). The Shannon-Weaver model was closely associated with behaviorism (Colombo, 2012). Psychologists suggested that human perception and memory were conceptualized in a capacity similar to the Shannon-Weaver model where sensory information is entered into receptors and then fed into perceptual analyzers. Subsequently, outputs are input into memory systems (Goldstein, 2011).

It is important to note that post–WWII, communications theorists were heavily influenced by the propagandistic efforts of both World Wars. Consequently, there was little distinction made among mass communication channels. It was virtually conclusive that advertising, media, and emergent technologies were viewed as nondistinctive opportunities to transmit messaging (MacDonald, Marsden, & Geist, 1980).

Another influential force in advertising was the prolific work of one of mass communications' founding fathers, Harold Lasswell (Berelson, 1964). Lasswell, a scholar and interdisciplinary social scientist, served as chief of the Experimental Division for the Study of War Time Communications during World War II. He analyzed propaganda media to identify mechanisms of persuasion and later applied his findings to behaviorism techniques and research (Pietila).

Lasswell proclaimed that mass communications demanded verifiable facts as a basis for generalizations rather than even the most brilliant individual insights. He advocated rigorous research to ensure predictability. When applied to advertising, such ideology affirmed its current direction. Moreover, it suggested that an emphasis be placed upon research and science as a way to ensure profits and sustainability.

More than 20 years after it was established as a full-fledged profession, advertising continued to achieve. Through prosperity and calamity, advertising revealed its ability to endure. By the close of World War II, advertising realized the goals it set out accomplish. With assurance, advertising was ready for even greater prosperity.

Colossal Expansion

As the years continued, advertising's success became increasingly apparent. For a sustained period of 15 years post–World War II, American advertising exploded. The gross total of advertising expenditures doubled in merely 5 years, from $2.9 billion in 1945 to $5.7 billion in 1950 (Fox, 1997).

Suburban expansion, construction, population growth, and new patterns of consumption led to a host of new products to advertise, specifically vehicles. In 1956 the Chevrolet advertising budget totaled $30.4 million, while the Ford budget totaled $25 million (Tungate, 2007). Demographics and geography fueled the largest advertising budgets to date.

Problem solving within the advertising industry had reached an apex. Work had become so plentiful that there was a healthy competition among advertising agencies (Tungate). JWT became the first advertising agency to exceed $100 million in billings, followed by BBDO, Young & Rubicam, and McCann Erickson. By 1957 all four agencies well surpassed $200 million in billings and they continued to soar.

The J. Walter Thompson agency achieved much of its success through innovation and fruitful relationships. For examples, JWT was the primary advertising agency of Kraft, a company well known for progressive advertising policies. For years they produced great advertising on a variety of media, including print, radio, and out of home. In 1947 the pair embarked on an unprecedented journey in television, the *Kraft Television Theater*. The program was the first regularly scheduled dramatic series on television. *Kraft Television Theater* was incredibly successful and is a highly revered classic of television's golden age.

Kraft Television Theater provided an incubator for creative experimentation. Thompson applied several of the techniques developed for radio and created new methods as well. The *Kraft Music Hall* radio program was relaxed, conversational, and featured celebrities, such as Bing Crosby. This tone was translated to television to ensure brand integrity. Thompson produced both programs and wrote much of the copy. His staff adapted the drama, hired directors, and cast performers. The program featured live adaptations of plays, familiar actors, and talented directors, including Stanley Quinn and Maury Holland. Commercials were eventually featured to demonstrate the uses of Kraft products. The "Kraft Hands" spots, which were among the first commercials to air on live television, featured no human faces and were considered nonintrusive and reassuring.

JWT provided significant models for the advertising industry and achieved great success. Prosperity was abundant during these times. As a result, their agency counterparts and competition did not trail far behind (Fox, 1997). As agencies grew in the years following the war, so did the expectations of clients. Accordingly, agencies expanded their staff and services. Agencies began to develop market research, merchandising, publicity, and public relations departments. Agencies were involved in product development, packaging, distribution, pricing, media, and more. Traffic departments became a welcomed addition to agencies as production, scheduling, and deadlines expanded advertising agencies in virtually all areas (Tungate). A trend of agency mergers began to develop as clients' demands increased. Consequently, in order to meet these demands, smaller agencies opted to merge with larger agencies as opposed to going under (Fox, 1997).

Postwar, the influence of psychology and interdisciplinary collaboration experienced a surge as well (Namba, 2002). The primary pioneer and agent behind this revival was Ernest Dichter, a strict Freudian psychologist and renowned advertising consultant. His success with campaigns for Chrysler and Ivory influenced many agencies to seek the collaborative efforts of psychologists who explored areas of the subconscious and catharsis. Accordingly, McCann Erickson became one of the first agencies to develop its psychological research staff, with many other agencies eagerly following this pattern (Fox, 1997).

In addition to research efforts, psychology was applied to visual and graphic communications (Benjafield, 2010). Deriving from experimental studies in Berlin, Gestalt psychology investigated mental processes and information organization in order to determine visual perception among humans. Gestalt was also utilized as a foundation for visual perception and media related industries. Its major processes involved providing a structure for effective communicative properties utilized in advertising, namely graphic design, figure-ground relationships, illustration, branding, and art direction (Arnheim, 1974).

In the 1950s the advertising industry experienced more growth and stability than ever before. Several major agencies expanded their operations overseas (Tungate). International accounts became increasingly lucrative as the American advertising empire ballooned. Moreover, the roles of practitioners evolved as advertising integrated further within society. For example, President Dwight Eisenhower appointed Neil McElroy as America's Secretary of Defense. McElroy was a successful advertising executive with Procter & Gamble and an expert in branding (Fox, 1997). Similarly, the Vatican named Saint Bernardino of Siena as the Patron Saint of Advertising.

Advertising became even further exalted through the introduction of television, which was referred to as the ultimate ad medium. Although television technology had been available since the 1930s, it gained tremendous popularity during the 1950s. Television had a long gestation period due to initial bureaucratic and technological issues. However, in 1949 the Federal Communications Commission (FCC) lifted its ban on new television stations. Commercials appeared in the 1950s and contributed a huge stream of new revenue for the advertising industry.

Soon after, methods to measure this medium emerged. In 1952 the Nielsen rating system for television was established and endorsed by the Advertising Research Foundation. Nielsen's machine-based ratings system became the primary way to measure the reach of television commercials (Moriarty, Mitchell, & Wells, 2012). Television became so successful that by the close of the decade nearly 90% of all households owned a television set (Hovland & Wolburg).

During the golden age of television, the lines between advertising agencies and networks began to blur (Fox, 1997). Preliminary television advertising was through sponsorship of entire programming. In its initial stages, interwoven relationships made it difficult to establish specific boundaries and standards. Consequently, abuse and corruption of television technology were a frequent occurrence. A system for selling individual spots and airtime was soon implemented. However, it was more profitable for television networks than for advertising agencies. Critics viewed this occurrence as yet another instance of American corporate monopoly (Fox, 1997).

The intense popularity of television during the 1950s marked a solidified relationship between advertising and yet another entity—the television media. Television joined newspaper, radio, and magazines as a linchpin in advertising media buys and national campaigns. In 1952 Rosser Reeves of the Ted Bates advertising agency used television to create advertisements for Republican Dwight D. Eisenhower during the presidential election of that year. Eisenhower won the election, thus, strengthening the relationship between advertising and media and politics.

By 1954 television became a leading medium for advertising. The year was also significant because during that same period, the Leo Burnett advertising agency introduced the iconic Marlboro Man as a character in its famous and highly successful campaign for Marlboro cigarettes (Fox, 1997). This commercial icon became renowned worldwide and a revered symbol of brand repositioning. In addition to its cultural appeal, the Marlboro Man campaign was economically profitable as well. According to *Ad Age* (Ad Age Advertising Century: Timeline, 1999), by 1955 sales for the product were at $5 billion, a 3,241% increase over the year prior.

The Marlboro Man became a powerful marketing tool for the Phillip Morris Company. Far surpassing the brand and product, the Marlboro Man icon resonated worldwide as a symbol of American culture and masculinity. Its widespread allure represented the successful usage of popular culture tropes and classic narratives within advertising. Figures, including heroes, cowboys, and princesses, were used to provide frameworks to attain universal cultural appeal.

Throughout the 1950s advertising experienced astounding breakthroughs. Advertising expenditures increased to unprecedented levels. By the close of the decade annual advertising industry billings grew from $1.3 billion to $6 billion. Advertising capitalized on a steady increase of product manufacturing, consumption, and population growth. In addition to adults, children and teens became valuable targeted audiences ("Ad Age Encyclopedia of Advertising," 2003).

However, in the midst of success, market penetration of television, government policies, and the proliferation of advertising sparked societal concerns. Research agendas were formulated to determine the impact of these factors (Fischoff, 2005). Due to many of its close partnerships, advertising experienced reprimands that corresponded with societal concerns. The inability to succinctly extricate disciplines foreshadowed impending challenges (Taylor).

Although the advertising industry experienced great prosperity throughout the 1950s, an air of tension loomed just below the surface (Fox, 1997). An underlying battle within the discipline between two distinct methods of the practice began to rise to the surface. There were those who believed that advertising should be more heavily reliant upon art and creativity and those who did not. Creative advocates suggested that a strong emphasis on artistry, copy, visual communication, and creative techniques would inspire consumers to purchase goods and services. Conversely, others embodied a more pragmatic approach to advertising. These groups of individuals believed advertising techniques should be deeply rooted in facts and research (Moriarty et al.).

Due to the strength of the underlined paradigm and its successful application abroad and during World War II, the latter approach prevailed. For much of the late 1950s, advertising focused on consumer influence, strict research, and the strategic use of modernity. Despite dissension within the discipline, timing, funding, problem solving, or events would not widely permit alternatives.

Shifts & Upheavals

The 1960s ushered in a period of significant societal change that led to shifts in numerous areas of advertising. America experienced a period of counterculture in which much of its ideology, politics, and contradictory behavior were challenged. Accordingly, this was reflected in advertising during these times. In the 1960s America was forced to examine its gender roles in society and urged to reform harmful environmental practices. While expanding American culture internationally through Peace Corps endeavors, the nation struggled with domestic unrest and ethnic inequalities. Upheaval experienced within society was reflected in advertising's workplace, audiences, and methods. Moreover, advocates of the creative approach would finally have an opportunity to advance to the forefront of advertising, thus, sparking an era known as the creative revolution (Tungate).

Critics of corporate mergers and big advertising agencies of the 1950s argued that their influence produced generic creative solutions that did not effectively connect with the American consumer. As a result, creative exploration increased. This exploration was aided by the prosperity of the 1950s, which facilitated the revenue for start-up boutiques and independent advertising agencies (Tungate). The decades of the 1960s and 1970s marked a period of resurgence for art, inspiration, and intuition in advertising.

This creative revolution was largely attributed to the works of three creative geniuses: Leo Burnett, David Ogilvy, and William Bernbach (Moriarty et al.). Burnett asserted:

> There is entirely too much dull advertising. Pages and pages of dull, stupid, and uninteresting copy that does not offer the reader anything in return for his time taken in reading it. A lot of this is just plain razzle-dazzle. We have been in thrall to the shibboleth of Bigness ... The creative men are the men of the hour. It is high time that they were given the respect that they deserve. (Fox, 1984, pp. 221–222)

Advertisements produced during these times implemented nonconventional methods. Creatives placed an emphasis of visual aesthetic and tone, which enhanced brand continuity. Memorable characters, white space, and savvy copywriting were used to attract audiences and differentiate brands. Advertisers learned that great success could be acquired through nontraditional practices.

When Levy's, a thriving local brand, wanted to expand its audience it sought the advertising expertise of Doyle Dane Bernbach (DDB). Subsequently, a campaign was developed to increase Levy's public acceptance. The strategy used literal interpretations in unexpected ways. Advertisements featured beautiful photographic imagery that represented America's changing demographics. DDB advertised Levy's Jewish rye bread with an Irish cop, Native American man, and African American boy. The iconic "You don't have to be Jewish to love Levy's Real Jewish Rye" campaign slogan became a highly revered success (Fox, 1984).

At first glance it may have appeared as if the advertising methods of Burnett, Ogilvy, and Bernbach were totally new due to their revolutionary impact. However, many of their approaches to advertising were influenced by methods of the past. As indicated through previous examples in advertising history, legacies were created through the circulation of knowledge within close-knit groups. A rich heritage developed as prominent figures handed down knowledge and methods to members within their communities. Groups that possessed the most affluence, whether through wealth or fruitful alliances, tended to dominate.

However, this did not indicate that other viable knowledge did not exist. In many cases, credible, yet less-prominent, concepts circulated within less powerful communities. When power shifted, as it did in the 1950s–1960s, multiple ideas ascended to the foreground. The prosperity of corporate ventures produced revenue, which was used to create independent advertising agencies. Thus, members of wider communities introduced concepts that were influenced by advertising leaders of the past. Accordingly, each of the leaders of the creative revolution of the 1960s was mentored under the works of a prominent figure in advertising history.

Burnett was mentored by advertising legend Theodore MacManus. Burnett developed an advertising approach known as the Chicago School of Advertising, which was heavily influenced by MacManus (Fox, 1997). Burnett's primary technique was known as inherent drama. In this technique, common narratives and cultural archetypes from American popular culture were used to create mythical characters that represented American cultural values. Burnett's most successful icons include the Jolly Green Giant, Tony the Tiger, the Pillsbury Doughboy, and the Marlboro Man (Fox, 1997).

Conversely, David Ogilvy proclaimed himself an advertising classicist. Ogilvy asserted that advertising once had a great period and he intended to return to it. Ogilvy fused the past styles of Raymond Rubicam, which were image driven, with the claim school of Albert Lasker. Ogilvy came to embody the ability to create strong brands through symbolism. Ogilvy's legacy includes product specific brands with information-rich claims, namely Rolls-Royce, Pepperidge Farm, and Guinness (Tungate).

Bernbach, considered the most innovative of the three giants, was a second-generation ad man as his father was an ad designer. Paul Rand, with whom he had worked early in his career, mentored Bernbach. Accordingly, he combined a keen sense of copy with design and innovative concepts. Bernbach is known for his focus on feelings, emotions, and intuition within advertising. Bernbach argued that there were numerous great technicians in advertising; however, persuasion is not a science but an art. In essence, advertising is the art of persuasion. Bernbach became widely known for his innovative Volkswagen campaign (Tungate). His legacy has been extremely influential in American popular culture and has contributed to advertising's social influence.

Although advertising was experiencing its own revolution, there were those who recognized that it was beginning to lag behind the world around them. Over the years, aspects of American advertising became notorious for depicting stereotypical and false representations of ethnic communities, including African Americans, Native Americans, and Jewish Americans. However, in the 1960s, social justice movements and technology pushed the barometer of change forward. Sharp contrasts between reality and advertising depictions became glaringly obvious. The United States was experiencing unprecedented social upheaval while advertising continued to place women and people of color in subservient positions. The Vietnam War, the civil rights era, the youth movement, and the sexual revolution challenged advertising's perpetual tendency to cater to marketplace norms and dated character values (Tungate).

During the 1960s critics suggested that advertising's outdated models contributed to cultural inferiority and subordinate positioning. There was an increased demand for accurate and dignified portrayals of people of color in advertising. Many believed this would most likely occur through legislation and inclusive workplace models. Advocacy groups requested that advertisers and advertising agencies incorporate positive imagery of black people into their campaigns and take steps toward a more diversified workforce. Members of the U.S. government supported these ideas as well. In the Kerner Report, which was issued by the National Advisory Commission on Civil Disorders in 1968, it was suggested that mass medium industries dominated by Caucasians would be unable to effectively communicate with diverse audiences (Wilson, Gutiérrez, & Chao, 2003).

The creative revolution marked a time when advertising began to become interactive with culture through connectivity. However, this line of reasoning was a distinct contradiction to established paradigms and theoretical communication models (Potter). Advertising communication was intended to be a linear process in which messages were transmitted and received. Communication was not intended to be circular nor was feedback sufficiently taken into account (Hovland & Wolburg). Advertising was intended to influence, not to react or respond to issues in society.

Despite its original intentions, advertising was now an institution of great prominence. Like citizens and advocacy groups, governing officials believed advertising had a responsibility to accurately reflect American consumers. Regardless of intentions or its original function, advertising was accountable.

As the creative revolution continued to interact with external factors, a realization of commonalities ensued. Mutual beliefs became increasingly apparent as the decade progressed. Moreover, generations of advertising professionals became aware of their role within American culture (Fox, 1997). Some members of the advertising community and the youth generation recognized that they shared a disdain for authority, strong visual sensibilities, and media curiosities. Author and advertising copywriter Edward Hannibal proclaimed:

> Just as the hippies, the kids and the Negroes were now beginning to raise hell with the country shaking up old farts and making them doubt themselves, advertising as a business was getting noise from the furnace room that it couldn't shut off any longer. (Fox, 1984, p. 270)

Additionally, the postwar baby boom was beginning to produce a significant impact within the workplace. As this generation matured, many sought careers in the advertising industry (Tungate). Once they joined the profession, this group infused their nonconventional views into their ads, seeking to make advertising serve the larger revolution. As consumers, this cohort was the first commercial generation accustomed to movies and television rather than magazines and books; therefore, many concluded that the consumers of this generation were primarily visual thinkers with reduced critical abilities. This led to concerns among seasoned advertising professionals. Critics argued that baby boomers' lack of traditional cultural roots led them to be exceedingly trend conscious, naïve, and obnoxious (Fox, 1997).

As the creative era came to a close, William Bernbach expressed his concern that consumers and advertising professionals were trying to be different for the sake of being different alone. Bernbach argued that young people working within agencies were mistaking the façade for the real thing. There was internal resistance toward the idea that advertising could be used as a catalyst to improve societal ills. Conflicts resulted between seasoned advertising professionals and recent members of the profession. Advertising executive Rosser Reeves predicted that advertising narcissists would wake up and big business would return as the immutable advertising law (Fox, 1984, p. 271).

Fortuitously, advertising's emphasis returned to hard sell, research, and science. After a decade that produced some of the most revered and innovative creative advertising solutions, communication strategies were altered. Increased competition, wealth, and a broadened societal scope propelled ideal consumers to the forefront of advertising's agenda (Coleman, 1998). The creative revolution resulted in a mastery of television, new approaches to print, and a renewed emphasis on photography and visual thinking.

Brands profited from creative success. Alka-Seltzer, in particular, provided unique creative examples that were often representative of larger shifts in culture and media. Scientific approaches gave way to strategic radio sponsorships and television media buys. The brand embraced newspaper, magazines, jingles, and developed promotional films for movie theaters. By 1951 Alka-Seltzer introduced "Speedy," an animated character conceived by Wade advertising agency. Speedy became a celebrated icon associated with the success of Alka-Seltzer. Speedy appeared in hundreds of print and television advertisements. By 1963 Alka-Seltzer achieved more success with the humorous creative approaches of Jack Tinker and Partners advertising agency (Archives, Alka-Seltzer Oral History).

In 1968 Alka-Seltzer partnered with the wildly successful creative genius at Doyle, Dane, and Bernbach (DDB) advertising agency. DDB produced popular television commercials for Alka-Seltzer. Although the advertisements received adoration and were considered to be hilarious by the public, Alka-Seltzer's market shares continued to fall. By the close of the 1960s, like many brands that had achieved marked success through creative solutions, Alka-Seltzer moved their $20 million account from DDB to Wells, Rich, and Greene advertising agency for economic promise, hard sell, research, and product emphasis (Fox, 1997).

In the 1960s changes abounded throughout advertising. In concurrence with the creative revolution and societal upheaval, psychology experienced major shifts as well. These shifts were predictable because of prior developments in computer modeling. During World War II, computer modeling supplied powerful structures for information representation and models that operate inner structures. Many of these ideas were successfully applied to human information processing (Miller, 2003). Subsequently, effective methods to study diverse topics ranging from pattern recognition, attention, categorization, and memory to reasoning, decision making, problem solving, and language developed (Bandura, 2005).

In addition to advances in information processing, several outside factors contributed to major shifts in psychology. The works of noted linguist Noam Chomsky had a profound impact. The central theme of Chomsky's work focused on mental structures and organization capabilities necessary for language competency. Chomsky developed syntactic structures that altered the intellectual landscape of psychology and fostered acceptance of cognitive sensibilities (Goldstein). Furthermore, Chomsky's 1959 annihilating review of B. F. Skinner's *Verbal Behavior* (1957) gained influential notoriety as one of the most significant documents in the history of cognitive psychology.

Advances in technology and scholarly research contributed to the eventual rejection of behaviorism as the dominant paradigm in favor of cognition. In what became known as the "cognitive revolution," it was concluded that behavior would be determined through inner processing. It was determined that the mind and mental processing were far too complex to be fully explained through predictable external factors (Goldstein).

Although the cognitive paradigm had yet to be articulated, this shift had a tremendous impact on advertising. Over the years advertising developed an essential relationship with psychology. The social sciences had become a valued and trusted tool within research and practice. Psychology informed and worked in tandem with many of advertising's communications models. Moreover, the advances that occurred in psychology inferred that advertising was applying dated methods. Furthermore, advances suggested that consumers did not process information as previously believed. As with culture and societal issues, advertising lagged behind.

Hills & Valleys

In the 1970s the monumental highs of the 1960s were met with equally significant lows. This decade represented an overarching compendium of patterns that came to embody the rhythmic flow of advertising. After a period of some of the most creative campaigns ever produced, emphasis returned to hard sell and research. Brand development was replaced by product differentiation. Teams and group development transitioned into hierarchal management structures. By far, one of the most impactful shifts occurred within audiences. The upheaval of the 1960s left many Americans skeptical and confused. Advertising embraced logic over intuition as a means to connect with increasingly distrustful and cautious consumers. Despite signs of decay, the unifying paradigm was still useful. Advertising's ability to harness technology and leverage strategic modernity proved to be reliable during one of advertising's most turbulent decades.

The economy fell into recession at the onset of the 1970s, which exacerbated conditions. As national traditions began to erode, so did those in advertising. Large agencies, including JWT, McCann, and Y&R, suffered enormous loss from domestic billings. Professional organization and economic procedures were reexamined for productivity. Moreover, formidable longstanding client relationships were dissolved. As advertisers desired assurance in an unstable economy, creative risks lost support. Advertisers left creative agencies in droves, which amounted to millions of dollars in declining revenue. Several creative agencies lost accounts, became conservative, or folded under confounding restrictions.

After a period of enormous success, many were bewildered by seemingly sudden and harsh change. However, others believed the calamity experienced during the 1970s was a residual impact from the previous decade. Critics advised advertising to readily embrace the American public in its entirety. As the world changed, advertising could no longer solely shape trends and exert influence. Advertising was now tasked to connect and respond with the needs and wants of consumers. As industries became increasingly consumer oriented, understanding the character of the American public became a critical issue for agencies and advertisers alike.

However, the ability to understand American consumers posed an enormous challenge. After experiencing significant upheaval the character of the nation itself was in flux. Disillusionment, spurred by the compounding assassinations of President John F. Kennedy and his brother, Robert F. Kennedy, and civil rights activists Medgar Evers, Malcolm X, and Martin Luther King Jr., was combined with the violence and aftermath of the Vietnam War. Moreover, the Watergate scandal—which ultimately brought down President Richard M. Nixon—deepened many Americans' levels of cynicism and distrust for tradition, bureaucracy, and government (Schulman, 2001).

Hence, the 1970s ushered in a unique phase of development for the entire nation. This period also revealed the changes that the American character underwent as the nation transitioned in a modern era. More so than previous generations, corporate agendas, peer groups, and mass media influenced Americans. As a result, the fabric of societal culture was significantly altered. A vast majority of Americans no longer relied upon traditional institutions for guidance. Instead, Americans began to turn inward or seek other factors for direction (Riesman, Glazer, & Denney, 1963). This, however, presented somewhat of a paradoxical conundrum for the nation. Since the post–WWII era, messaging that encouraged conformity and individuality simultaneously influenced Americans, particularly youth. As a result, there were contradictions within consumer behavior.

Corporate influence contributed to messages of conformity. In order to cultivate desirable market behavior, consumption and relatability were emphasized as favorable traits. Mass communication, including advertising, stimulated ideals and habits among American consumers. Young children became elevated opinion leaders and were taught to cultivate consumption patterns with their parents and within their families. Teenagers were identified as cultural tastemakers. Although such strategies created brand loyalties and stimulated purchasing behavior, they also contributed to confusion. Boundaries surrounding consumer identity and the responsibilities of mass media were blurred (Riesman, Glazer, & Denney).

The influence of corporate institutions and consumer culture represented a broad, yet pervasive, cultural shift. Extraordinary transformations in social values were further complicated by experiential factors that promoted individuality. For example, Americans prided themselves on the spirit of individualism within the political arena, demanding the right to freedom of speech and religion. However, when confronted with social issues, the nation primarily responded in hypocrisy and conformity (Schulman). Thus, the 1960s produced radical individualism that extended from the political arena to personal lifestyle. Previously held traditional values were no longer congruent with the needs and wants of the vast majority of the generation coming of age during the 1970s. A greater emphasis on self-reliance developed. Through experience, Americans learned to be skeptical and distrustful of institutions. Americans learned that they could not solely rely on institutions, namely government, religion, and family, to provide an ideal way of life.

It was assumed that institutions were the blame for societal inequities and therefore responsible for rectification. Along with most American institutions, the public grew dissatisfied, disinterested, and skeptical of advertising as well. According to a 1970s Gallup poll, American citizens ranked advertising executives dead last when asked to rate the honesty and ethical standards of working professionals. Thus, regulation, deregulation, and its resultant tension became common themes in the 1970s.

As a societal institution, the public held advertising to obligatory responsibilities. Precedents associated with the Civil Rights Act (1964, 1965), Voting Rights Act (1968), and *Roe v. Wade* (1973) were mimicked throughout the advertising industry. Regulation pushed advertising toward ethnic and gender pluralism and elevated imagery. Advertising and its effects became the subject of increased scholarly study and scrutiny. Moreover, the Federal Trade Commission (FTC) held the advertising industry to an unprecedented level of honesty and disclosure. This led to a diversified workforce, increased ethnic representation in advertising, varied creative solutions, and awkward discontent. Once a supporter of advertising regulation, advertising executive David Ogilvy began to reconsider his position.

Ogilvy stated, "I guess we deserved some of this; we failed to regulate ourselves. But we are now overregulated. The whole thing has become a farce" (Fox, 1984, p. 319). According to an AAAA study conducted from 1970 to 1974, minority employment within 38 of advertising's top agencies increased from 8.9% to 9.9%. However in 1975, the AAAA reported that African Americans were experiencing increased tensions within agencies and being denied positions of executive leadership. In fact, African Americans within leadership positions at advertising agencies decreased from 4.6% to 4.2% from 1970 to 1974 (Fox, 1997). Winthrop Jordan, an African American adman, declared that integration efforts had led to a degree of self-deception within the advertising industry (Jordan, 1975). Accordingly, many African Americans filed employment discrimination charges against most of the top advertising agencies during the 1970s (Fox).

Advertising experienced upheaval within the area of gender norms as well. Although there was a female professional prescience in advertising, pressure from inside and outside of the industry forced gender roles and female imagery to improve. There was an applied understanding that women's interests exceeded feminine products. Nonetheless, like ethnic minorities, few women held executive positions within the industry. Advertising giant Mary Wells recognized this inequality and acknowledged that she knew of several women who were qualified to perform at executive levels in advertising. However, she herself did not readily advocate gender progression. Wells asserted that male clients would not tolerate women at executive levels in advertising and that chauvinism prevented men from dealing comfortably with women of authority (Fox, 1997).

Despite opposition, there was some progress. In 1973 Reva Korda became a creative director at Ogilvy and Mather. Additionally, after a series of sexual discrimination charges, representation of women in professional jobs in advertising rose from 40% to 57% by 1982. Moreover, although they folded due to financial pressures, the 1970s witnessed the acquisition of major general market accounts by African American–owned agencies, among them John Falls Shop and Zebra Associates (Fox, 1997).

Although both America and advertising had begun to evolve, advertising's inextricable role within American consumer capitalism retarded its development. In addition to social regulation, areas of particular impact within advertising were its economic structure and creative practices, all of which were significant to both the internal and external environment within advertising, and which foreshadowed a major forthcoming transition and eventual crisis. As predicated by existing advertising cycle norms, the industry shifted from creative prominence to a renewed emphasis on hard sell, science, and research. Yet, there were several factors that made advertising practices in the 1970s distinct.

During this period, there was a significant shift in advertising's economic structure. Creative boutiques could not withstand the economic pressures of recession and gave way to new practices and corporate mergers. However, unlike the economic practices that occurred in previous advertising cycles, the 1970s implemented public ownership to combat financial pressure. At the beginning of the decade, half of all top agencies had gone public, indicating the beginning of a trend. Although public ownership often supplied the cash reserves agencies needed to survive the turmoil associated with recession, it left the advertising industry obligated to the needs and wants of stockholders.

Advertising executive George Lois declared that public ownership was the catalyst for destroying partnerships (Fox, 1997). Consequently, several agencies began to repurchase their common stock and return to private ownership near the middle of the decade. However, this insight was determined after economic practices had been altered. Therefore, many of the remaining agencies were unable to regain private ownership and sought to embrace the changing economic environment.

Additionally, the mergers that occurred in the 1970s were also distinct. Increased business and marketing leadership throughout the advertising industry encouraged free enterprise and expansion. In the 1970s advertising executives had visions of "total agencies" that would expand advertising expertise and dominion. More profitable agencies took steps to achieve these goals, which resulted in a growing trend throughout the advertising industry. Numerous regional agencies and specialty boutiques, facilitated by the creative revolution of the 1960s, were targeted by larger, more successful advertising agencies. Economic instability, coupled with the prevalence of merging patterns, resulted in a serious decline of independent advertising agencies nationwide. In fact, no major independent advertising agencies existed on the entire West Coast by the close of the decade. James Heekin, an advertising executive of a small consulting firm, proclaimed that small agencies of that period had lost their entrepreneurial confidence. "They see God on the side of the big battalions. They don't want to be the last guy on the beach, so they figure they should get aboard," he stated (Fox, 1984, p. 327).

Within the realization of the vision for total agencies, executives sought to primarily acquire burgeoning independent enterprises rather than crippled operations in economic distress. Moreover, the advertising industry began to broaden its range of services and exceed seemingly infinite client demands. Young and Rubicam (Y&R) CEO and Chairman Edward Ney established an industry precedent as he worked toward achieving what he termed "a wide horizontal range of commercial persuasion" (Fox, 1984, p. 326). Ney bought advertising agencies and boutiques that specialized in a range of expertise from health care and direct marketing to retailing and sales promotions. By the close of the decade Y&R skyrocketed as an international leader, reaching billings of more than $300 million worldwide. Many within the advertising industry began to model these practices and resultant success. The 1970s also witnessed the acquisition of the three largest public relations firms: J. Walter Thompson, Y&R and Foote, and Cone & Belding (Fox, 1997).

Internal upheaval and instability within the advertising industry forced many standard practices to shift in accordance with acquisitions, new media, expanded practitioner responsibilities, and fiscal procedures. Traditionally, advertising agencies relied on a system of compensation that yielded 15% commission on media billings. However, during the 1970s there were struggles among agencies, partners, and clients regarding this economic practice. To the benefit of high-end clients, a fee system was implemented throughout the advertising industry. Although the fee system offered agencies more definite forms of payment, this system primarily favored agencies that advertised new products. The traditional commission system rewarded effective advertising for most products as well as longstanding agreements. Moreover, the commission was the preferred norm throughout the advertising industry. Controversies surrounding the fee system resulted in intense friction throughout advertising. After much debate, nearly three-fourths of all agencies returned to the traditional practice of commission by the close of the decade (Fox, 1997).

Distinct shifts and occurrences during the 1970s also took place within advertising creativity. Although the technology, strategies, and procedures developed as a result of the creative revolution during the 1960s were formidable, hard sell and management practices prevailed during the tighter economic climate of the 1970s. Creative awards and acclaim no longer guaranteed financial success. Under the direction and leadership of MBAs, advertising agencies began to expand in the areas of marketing, pricing, distribution, and packaging.

The dominance of business within advertising during the 1970s contributed to the widespread usage of positioning as an advertising technique. Positioning addressed the needs of expanding markets through comparative strategy. Although it was a resurrected technique from the hard-sell approach implemented during the 1950s, Al Ries and Jack Trout, of the Ries, Cappiello, Colwell Agency, contributed to its renewed popularity and widespread application. Within their conceptualization of positioning, product differentiation and branding were emphasized. Moreover, it offered a creative way to emphasize specific information and facts without relying on puffery and generalities. This was highly welcomed during a time of intense regulation. Ries went on to co-author along with his daughter, Laura Ries, *The Fall of Advertising and the Rise of PR* (2002) in response to additional shifts in the industry. In it, Ries advocates the use of experiential interactivity within advertising.

Additional shifts in advertising were made through paring down agencies, the adoption of creative directors, and the expansion of international departments. With the decline of domestic commercial opportunities, American advertising agencies sought to service multinational clients, including Ford, Coca-Cola, and Procter & Gamble. Hence, American advertising agencies began a modern expansion into the European and Australian international markets.

Within these uncharted territories, advertising used giants in the industry as models for success. In addition to burgeoning international development, Y& R became a standard for creative modeling. Creative director Alex Kroll led the agency with astonishing creative growth in a time of economic turmoil. Kroll insisted that advertising must remain manic about creative discipline. However, in contrast to the creative revolution of the 1960s, Kroll declared that discipline to be just as vital as creative. He maintained that creativity is tactical, stating, "Creativity should be measured by the cold, gritty eye of the marketplace, not by the vibes acquired in a screening room" (Fox, 1997). Kroll established the precedent that advertising creativity be judged by the consumer takeaway and not peer applause. Thus began the shift toward consumer-driven advertising.

As more of an emphasis was placed on consumers, technology proved to be a critical tool for advertising. Geodemographic clustering systems technology was introduced in the 1970s. This technology facilitated advertising's ability to shape and meet the needs and wants of consumers in a more sophisticated manner. Advertisers assumed that consumers who resided in similar neighborhoods or postal codes zones shared demographics and lifestyles. Hence, the ability to obtain this information allowed advertisers to determine consumer purchasing power and marketplace behavior. Geodemographic clustering technology presented statistical analyses and databases at more defined levels than previously available within the national and international mass markets. Advertisers used the information to label and characterize consumers in illustrative categories including bohemian mix, suburban pioneers, and upper crust. Additionally, geodemographic clustering technology assisted advertisers in making vital marketing communications (marcom) decisions, including selecting geographic locales for narrowcasting media advertisements and identifying audiences for direct marketing campaigns (Shimp, 2010).

Despite instabilities spurred by economic and social agitation, areas of technology helped sustain advertising during the distressed 1970s era. Moreover, color television, FM radio, and new production techniques were developed in the mid-1970s, which rejuvenated media, achieved commercial success, and fortified global influence. Additionally, special effects helped catapult the film industry to even greater prominence. Consequently, the 1970s, which featured the major profitable blockbuster films, provided opportunities for product placement and branding for the advertising industry (Schulman).

The 1970s period also became known as a fashion era, with fabric and design technology spurring the production of synthetic and natural fabrics in a variety of patterns and colors. The advertisement industry played a dominant role in this development, particularly in the campaign promotion of denim jeans. Although previously associated with the Beat Generation subculture, advertising elevated denim jeans to a widely accepted and highly revered quintessential symbol within American popular culture. Denim jeans personified the rebellious individualism of the 1970s and embodied labor, leisure, style, and comfort, regardless of class (Frum, 2000). Moreover, new technology accelerated production, enhancing fashion cycles. Advertising used effective strategies, branding, and positioning techniques to increase product consumption, thus, complimenting cultural and economic shifts of the 1970s associated with two-income households, diverse audiences, and increased female earning power. Additionally, advertising relied on its sustaining principle of modernity through the cycle of product obsolescence to achieve continued success (Frum).

Despite obvious success through technological advances, global changes within the economy led to increased fragmentation and significant unrest. Contributing to these challenges was a growing number of Americans who retreated into solitude, attributing their discontent to growing pressures from the pace of society, environmental concerns, monetary demands, credit woes, time restraints, and the relentless proliferation of technology and mass media (Fox, 1997).

Following the lead of consumers, advertising echoed society's individualistic sentiments by communicating that self-indulgence was not only an appropriate but also a critical element for human development. Personalization was further conveyed through advertising campaigns, which introduced new technologies, including the VCR, video rental, cassette tape, and cable television. Imminent fragmentation characterized the close of what was coined the "Me decade" in the 1970s (Hovland & Wolburg).

· 4 ·
DRIFT & DECLINE

Nearing the onset of yet another phase of development, advertising began to drift toward decline in the 1980s. Although this decade birthed several of contemporary advertising's most celebrated brands and influential achievements, it was also a precursor to crisis. While there were several factors that contributed to decline, significant events involving corporate influence and cultural disconnect were extremely detrimental.

Economic prosperity experienced in the late 1970s led to increased corporate involvement in advertising. Although this formidable alliance produced seemingly advantageous results—from 1975 to 1984 advertising experienced the largest boom it had ever known—it contributed to industry tension (Tungate, 2007). One main area of tension stemmed from corporate comingling, which led to blurred boundaries and growing complexity. Another area of industry tension derived from a heavy emphasis on research.

Tension mounted as advertising attempted to predict and control outcomes that could not be easily influenced or measured. In the 1980s audiences became increasingly resistant to previous methods and strategies, which eventually contributed to cultural disconnect. Conditions worsened as maturing markets became saturated and consumer behavior became progressively erratic. In response, advertising became even more emphatic about research and testing. Scientific methods were viewed as ways to satisfy clients and ensure successful outcomes.

As corporate influence continued to expand research efforts, it eventually became entrenched in the creative process. Advertising creative professionals were resistant to research demands. Following the creative revolution of the 1960s, advertising creative professionals acquired new knowledge, diverse experiences, and success. They could not easily be stifled or coaxed into the restrictive creative process that accompanied research dominance. Tension mounted, which foreshadowed impending factions.

As with previous cycles of advertising, the profession demonstrated its resilience through the ability to rely on its trusted paradigm. Accordingly, media, technology, and culture became a valued resource for creativity and nurturing collaborations during the 1980s. Advances in media and technology spurred cable television, which provided a breadth of advantageous opportunities for advertising, as well as corporate comingling. Moreover, the 1980s revealed a surge of fresh creativity that swept the nation. Culture was expressed in unique ways and provided inspiration for advertising through popular culture, hip-hop, and ethnic communities.

Popular culture produced megastars who were used as powerful influencers among audiences. One of the most valuable models for this success was attained through Michael Jackson's *Thriller*. This creative collaboration established a precedent that would lead to blockbuster advertising and celebrity influence. Hence, the 1980s were filled with major campaigns that featured pop culture celebrities and brand affiliations. Some of the most notable campaigns included Pepsi and Michael Jackson, Jell-O and Bill Cosby, and Nike and Michael Jordan (Kern-Foxworth, 1994).

In addition to popular culture and celebrity influence, advertising began to tap into hip-hop culture and ethnic communities for fresh ways to influence consumers. The success of Nike's affiliation with Michael Jordan, through brand integration, culture, and lifestyle, offered a promising direction for advertising to pursue. During these times advertising's ability to influence consumers began to diminish, particularly among youth audiences. The repeated attempts of advertising to establish relationships and cultivate loyal consumers became increasingly futile. Skeptical members of youth audiences perceived the use of culture by advertisers as disingenuous and ineffective (Kern-Foxworth). Accordingly, gaps between advertising and consumers widened.

Despite groundbreaking achievements, waves of prosperity during the 1980s were short lived. By the close of the decade tension that loomed below the surface of advertising began to mount. During this drift toward decline advertising was set on the path to collide with triadic convergence. Corporate influence and cultural disconnect contributed to vulnerabilities in advertising that impeded its ability to withstand the impact. In the wake of a potential paradigm shift associated with this collision, the explication of change becomes a central focus.

Influence & Obligations

The 1980s marked a period of perilous decline for advertising. While the onset of the decade revealed some of advertising's greatest and most lucrative accomplishments, its close was a stark contrast. Advertising was engulfed in crisis, entangled in obligations, and fragmented by factions. Although corporate influence produced advantageous results, it set advertising on a path filled with pitfalls.

The dominant presence of business in advertising, which spanned the latter half of the 1970s well into the beginning of the 1980s, led to seemingly beneficial outcomes. Free enterprise provided a respite for advertising during tumultuous economic and social times. From 1975 to 1984 the advertising industry experienced the largest boom it had ever known. Expenditures on advertising rose an average of 13% each year (Tungate). Business mergers and takeovers, which resulted in multibillion dollar growth and global expansion, characterized these times. The vast majority of wealth stemmed from multinational clients, many of whom had grandiose expectations and sought big agencies to fulfill them. Although conglomeration was economically profitable, it was often at the expense of smaller agencies and eventually the advertising industry itself.

In addition to demands associated with consumer culture and market maturation, conglomeration placed a substantial amount of pressure on advertising. Corporate influence created expectations that could not be easily met. For example, in seeking to understand audiences, technology, and process, challenges associated with mergers and acquisitions were largely ignored. Rigorous testing became a crutch for clients and executives.

Critics of rigorous testing believed it exacerbated tension and stifled productivity within advertising. Conversely, advocates believed research created a sound scientific approach that would satisfy clients and offset many of the challenges experienced within advertising during this time.

By the 1980s America had been deeply entrenched in consumer culture for nearly a century. Not only was the public growing tired of advertising, domestic markets were demonstrating signs of maturation and product saturation. A severe problem facing advertising practitioners was that products and goods had drastically changed. Goods had now become products of profuse parity. Within every product category there were four to five major companies producing the exact same thing. Advertising executive Rosser Reeves affirmed, "Our problem is a client comes into my office and throws two newly minted half dollars onto my desk and says, mine is the one on the left. You prove it's better" ("Advertising: A History from 1980–Current Day," 2009, p. 2). There were also inconsistencies involving product positioning and value. Clients believed their products provided superior experiences, while consumers did not.

To circumvent product emphasis, advertising placed a renewed emphasis on branding as a creative method. Hal Riney, creative director at BBDO, once asserted that brands needed to differentiate themselves or they would die. "Most of the time it doesn't seem like the facts have done me any good. It seems that there is already someone using the exact same ones" (Raine, 2008, p. 2). Despite the creative push to implement brand differentiation, there was resistance. Advertising creative professionals asserted that effective branding required time and development. However, this demand was often incongruent with the needs, wants, or schedules of clients. Moreover, brand persona was a relatively new approach. During these times brand personality was not considered a key distinguishable feature and was deemed esoteric by clients and business executives (Moriarty, Mitchell, & Wells, 2012).

An outgrowth of this vexing internal dilemma was segmenting or determining which products could be sold to which audiences. Advertising agencies determined that it was impossible to sell all products to all audiences. Instead, they began to reexamine what they were promoting in order to best determine who was going to buy their products. Once this determination was made, advertisements were placed in locations that were ideal for their corresponding audiences. Moreover, advertisements were designed to look, communicate, and identify with the needs and wants of these designated audiences. It became prevalent that advertisement messaging and aesthetic were created to focus on specific audiences. As a result, research became a deeply entrenched and necessary component within advertising creativity.

In addition to creative research requirements, client demands for research often placed increased pressure on advertising business executives and account management. Client research demands were often looked at unfavorably because they were restrictive, impacted media buys, and became a costly expense. Many advertising agencies were forced to transition from traditional radio and television advertising toward sales promotion techniques such as rebates, coupons, and sweepstakes. Although not always as impactful, these tactics provided measurable proof of sales that clients demanded (Moriarty et al., 2012). Moreover, consumer-driven research became increasingly expensive, particularly the employment or contraction of advertising researchers. Independent agencies were under enormous strain as they sought to meet the demands of evolving clients and the marketplace (Sacks, 2010).

The desire for precision-driven research contributed to restrictive creative methods. Focus groups, cognitive patterns, and copy testing became a staple within advertising creative departments. Art was heavily shifting toward science once more. Decisions that were made based on experience, judgment, and instinct were now made based on research and testing. The struggle over the direction of creative in the 1980s marked significant internal strife within advertising.

Creative Oasis

As in previous cycles, advertising turned its attention to its relationship with media to offset challenges experienced within the profession. Such attention had previously proven to be not only reliable but also profitable. Moreover, in the 1980s, there was a robust energy surrounding media that provided new inspiration, primarily due to the renewed zeal and spirit of popular culture. However, media, technology, and culture were beginning to shift, thus providing an unstable foundation upon which advertising could rely. Nevertheless, a short wave of nurturing collaborations provided an oasis for creative expression in advertising.

The impact of American consumerism helped create a multitude of status seekers. Subsequently, a unique culture developed that celebrated lifestyle and expression. In the 1980s labels and logos were abounding. Deregulation produced new billionaires and widened economic and social gaps within the American public. The population of the Sun Belt had risen to exceed that of the industrial regions of the Northeast and Midwest. Urban areas encountered a contraction of their economic base as enterprises also migrated to suburban areas for lower taxes and loosely regulated business environments. In turn, America's major urban areas experienced population declines, increases in unemployment, and demands for social services. Moreover, America experienced formidable shifts in its political climate, strengthened levels of conservatism, and a major drug epidemic (Batchelor & Stoddart, 2007). A fresh and eclectic cultural environment was formulating in America, one that would proliferate diverse forms of self-expression.

The 1980s were a huge decade for art and creativity. A wave of fresh artists depicted culture in unique ways and provided inspiration for advertising. Keith Haring, Andy Warhol, and Jean-Michel Basquiat emerged as leaders within the art world and became highly influential popular culture icons. New forms of music and fashion also provided stimulus. Additionally, advertising found new industries to popularize, namely fast food, candy, sports, and toys. Moreover, collectibles and paraphernalia became enormously popular. Advertising recognized lucrative opportunities within brand and product extensions and began furthering their relationships with television and film through product placement (Batchelor & Stoddart).

In addition to the emphasis on art and creativity, the 1980s era is known as the golden age of television advertising. The industry soared to new heights with groundbreaking broadcast programming, as shows such as *Dallas*, *Dynasty*, and *The Cosby Show* commanded upward of $400,000—the highest advertising rate for a series program known to that date. Only the Super Bowl rivaled this precedent for a 30-second commercial (Shiver, 1986). Moreover, advertising experienced rejuvenation through cable television. As cable and satellite technology became increasingly popular, advertising launched its own cable channels and cable formats, among them QVC, Home Shopping Network, Shop TV Canada, and infomercials. As a signifier of media intermingling, advertising created 15-second commercial spots to accommodate the brevity associated with cable television and the demands of clients and networks (Batchelor & Stoddart).

Unlike other advances in media technology during this time, cable television provided a breadth of advantageous opportunities for advertising. Cable television supplied an effective way to reach consumers by harnessing new directions spurred by segmenting and audience-driven tactics. Moreover, conglomeration supplied large agencies with plentiful budgets and new associations. There was extreme promise for profit acquired through successfully selling brands and products through lifestyle integration. Cable television provided an opportunity to do just that. The proliferating number of channels offered advertisers access to a host of targeted audiences, including children (Nickelodeon), sports fans (ESPN), movie enthusiasts (HBO and Showtime), women (Lifetime), news watchers (CNN), families (Disney), and more.

One of the most fruitful and pioneering partnerships came through advertising's relationship with MTV (Batchelor & Stoddart).

Music Television (MTV) launched as an American cable television channel on August 1, 1981. During this time Viacom, an American media conglomerate, owned MTV. The expressed intention of MTV was to assist music artists in gaining exposure in an increasingly visually driven culture. The format was a continuous loop of music videos driven by on-air hosts, known as video jockeys (VJs) (McGrath, 1996). The target audience for MTV was the youth culture market. American youth comprised the primary targeted commercial market for advertisers and manufacturers since the 1950s (Pasek, Kenski, Romer, & Jamieson, 2006). MTV created a historic opportunity for advertising. Advertising increased its potential to expand its influence on youth culture and lifestyle.

George Lois, art director, designer, author, and advertising creative icon, is credited with catapulting the MTV network to a level of unprecedented achievement with his emergency trade advertising campaign, I Want My MTV. Prior to the launch of the campaign, MTV was struggling to gain acceptance within the cable community. A large majority of cable operators refused to carry the MTV network. MTV executives wanted to use advertising to connect with their potential fan base in hopes that their demands would influence distribution within the cable community (McGrath).

In the 1980s it was increasingly obvious that product was no longer king and status, esteem, and affiliation were becoming more and more significant. Lois combined the bold and demanding spirit of American youth culture with the zeitgeist of the 1980s. The I Want My MTV advertising campaign featured a heavy emphasis on visual identity, branding, and rock superstars. Music celebrities were incorporated into the campaign to increase the credibility of the MTV brand among American youth. The campaign's commercial ignited a firestorm of frenzy. Lois (2013) affirmed:

> The cincher in each commercial was this windup sequence as a voice over proclaims: If you don't get MTV in the area where you live call the operator and say ... We then cut to Mick Jagger, who bellows into the telephone, I want my MTV! In each city, thousands called moments after viewing the commercial and screamed, I want my MTV! (p. 1)

Peter Townshend and Pat Benatar were also featured in commercials that appeared in heavy rotation in each market. Within months, MTV was in 80% of American households and advertisers considered the network as a "must buy" in order to reach viewers ages 14–28. Moreover, rock superstars around the globe brokered deals with advertisers in order to be a part of this highly influential pop culture campaign. The roster included major recording artists across the industry, including Stevie Nicks, Madonna, Sting, Van Halen, Cyndi Lauper, ZZ Top, Hall & Oates, and The Cars. After the success of MTV's advertising campaign, *Time* magazine declared that MTV was the most spectacular pop culture phenomenon since the advent of cable television and arguably since the invention of television itself (McGrath).

The success of MTV facilitated the expansion of its format to include not only music but also youth-oriented soap operas, animated series, talk shows, comedies, game shows, and reality television. With its success, MTV took major steps toward becoming the authority of youth popular culture.

Groundbreaking Precedent

One of the most salient steps in the development of contemporary youth culture was also a distinct reflection of the evolution of American pop culture during the 1980s. MTV expanded its music genres to include not just rock but also pop, rap, electronica, grunge, and hard rock. This was made possible by the groundbreaking historical achievements of recording artist Michael Jackson, subsequently deemed the King of Pop.

Michael Jackson (1958–2009), recording artist, entertainer, actor, and businessman, is distinguished as the most successful global entertainer of all time by Guinness World Records. Jackson's influence on pop culture spanned more than four decades. In 1982, just off of the success of the I Want My MTV ad campaign, Michael Jackson was a megastar on the verge of astounding commercial success. In that year, Jackson released his solo album *Thriller*, which led to enormous breakthroughs in the entertainment industry, popular culture, and advertising.

During this time, MTV was extremely rigid, expressing its musical format and targeted audience as rock only. This focus left many artists of color out of the rotation of the MTV network. After protests from celebrities, including rock recording artists David Bowie and Rick James, as well as accusations of racism and alleged threats to pull advertising and content, MTV relented. Contributing to the pressure was a highly publicized rumor that circulated throughout the industry. This story alleged that Walter Yetnikoff, group president of CBS Records, which was the parent company of Epic Records—Michael Jackson's music label—threatened to remove all CBS advertising and content from MTV unless it aired Michael Jackson on its network (Taraborrelli, 2009).

In March 1983, one week after the song hit No. 1 on Billboard's Hot 100, Michael Jackson's "Billie Jean" video debuted on MTV and is credited with breaking down racial barriers and transforming the relationship between music and media into an art form and promotional tool. Moreover, this revolutionary short film garnered mainstream attention for MTV and is considered one of the most revolutionary songs in the history of pop music. The debut of "Billie Jean" on MTV marked the beginning of a new invigoration in popular culture. The acclaim surrounding "Billie Jean" paved the way for explosive growth and subsequent intermingling.

Conglomerates capitalized on this unprecedented success. Momentum was utilized to propel the music video for the music single "Thriller" to unparalleled achievement. "Michael Jackson's Thriller," also considered a short film, is a 14-minute array of visually arresting and iconic imagery. The debut of the film is referred to as a watershed moment in history due to its seminal merging of filmmaking, music, audiences, and popular culture (Taraborrelli). *Making Michael Jackson's Thriller*, a documentary showcasing the production of the video, was released in tandem. Both were shown in heavy rotation at the request of demanding viewers on MTV. "Thriller" not only was revered by the American public but also by critics, the global entertainment enterprise, and the world at large.

At the time of the release of *Thriller* in 1982, the music industry was in decline. Between 1980 and 1982, record shipments were down by 50 million units, and sales had declined by more than 15% industrywide (Greenberg, 2012). Bruce Swedien, the record's music engineer, asserts that prior to production of the album acclaimed producer Quincy Jones informed those associated with the record that it was a benchmark that would save the music industry.

In the 1970s technological shifts dismantled the core of the mass audience upon which pop culture and media depended, ending AM radio's historical prevalence. At the close of the 1970s, 50.1% of radio listeners were tuned to FM. By 1982, FM commanded 70% of the listening audience. Moreover, FM radio reached 84% of the 12–24-year-old demographic (Greenberg). Consequently, each audience segment had only limited exposure to diverse entertainment. Billboard columnist Mike Harrison notes that "there was no longer an exclusive top 40 anything, but rather an ever-changing multitude of top 40s, depending upon the genre (which research would be focused on)."

He added that "those who enjoy a-little-bit-of-this-and-a-little-bit-of-that constitute a minority" (Greenberg, p. 1).

Precision targeting of audiences forced media to avoid anything that fell outside their target audience's narrowly defined tastes. Failure to do this often incited audiences to tune out (Greenberg). This plagued both the television and radio industries, as remote control technology became increasingly prevalent in the 1980s and increased fragmentation drained a great deal of fervor from pop culture.

There was no longer significant cross-fertilization between content and culture. A seemingly impenetrable wall had been erected between audiences. Hence, the magnitude of *Thriller*'s success was virtually immeasurable, ultimately leading to the disintegration of traditional racial barriers and negative aspects of audience segmentation (Taraborrelli). Music executive Steven Greenberg insists that the success of *Thriller* taught the industries that "the right star, with the right product, and the right technological environment, always has the ability to move us and to unite us all" (Greenberg, p. 7). The coordinated rollout of the film *Making Michael Jackson's Thriller* was fortified with invigorated features of popular culture and strategic elements of an IMC advertising campaign. These efforts would set new precedents and establish standards that the industry would soon follow.

Like advertising, "Thriller" was the result of masterminds from varied industries. A team of remarkable leaders whose talents ranged from business and technology to media and creative expertise, devised a virtually infallible product that would secure the affinity of a widespread audience. Advertising strategies that leveraged borrowed interest through esteem and affiliation were applied within the campaign. Celebrities were incorporated that were popular among differing audience segments. "The Girl Is Mine," Jackson's duet with Paul McCartney, was the first music single. McCartney, a global superstar and member of the Beatles—the most commercially successful and critically acclaimed group in the history of popular music—was an icon among a pop culture audience. "The Girl Is Mine" was followed by the hit single and short film "Billie Jean," which harnessed new technologies and media. The single exposed Michael Jackson to an entirely new audience, which was primarily composed of suburban youth. "Billie Jean" was followed by "Beat It," which featured Eddie Van Halen, lead guitarist and co-founder of the seminal hard rock band, Van Halen. Yet again, Jackson was exposed to an entirely different audience through this association. Moreover, the video choreography for "Beat It" was revered as an urban rendition of *West Side Story*, a popular American classic film. The video soon became viral and was subsequently inducted into the Music Video Producer's Hall of Fame (Taraborrelli).

With a product that appealed to diverse audience segments, conglomerates secured airplay on a variety of radio stations. Moreover, Jackson's videos were in high rotation on cable television, the new media. On May 16, 1983, with "Beat It" at No. 1 on the music charts and "Billie Jean" still among the Top 10, another component was strategically added that would integrate media selections, audiences, and bolster interactivity with audiences. Jackson debuted the moonwalk dance move on the *Motown 25: Yesterday, Today, Forever* 25th anniversary TV special, which aired on NBC. The premiere of this event on broadcast television was a strategic decision. The majority of Americans had yet to acquire cable television. Consequently, the use of NBC solidified yet another audience. Buzz, generated through radio airplay, heavy rotation of videos utilizing new media, and teaser snippets featured on broadcast television, elicited a mass and diverse audience for Jackson's performance. It also created intrigue for Jackson's upcoming single titled "Thriller." Forty-seven million Americans tuned in to see Michael Jackson. The moonwalk became Jackson's signature dance move and has been emulated incessantly across the globe. Jackson's performance created a viral sensation and bolstered the growing frenzy surrounding "Thriller."

Pop culture has always thrived on mass excitement. This synergistic campaign titillated not only the American public but also music enthusiasts worldwide. On November 12, 1983, nationwide, and January 23, 1984, globally, the "Thriller" single was released to a captivated public. The success of "Thriller" catapulted into the stratosphere primarily due to its music video accompaniment. With the aid of new media and creative genius, conglomerates attained the ability to mesmerize the public.

Elements of pop culture and nostalgia were incorporated to fortify the success of "Thriller" and ensure its duration as a commodity for years to come. The video for the single was a short film that paid homage to 1950s classic horror films, such as the cult classic *An American Werewolf in London*, directed by John Landis, who also directed "Thriller." Costarring with Jackson was former *Playboy* centerfold and model Ola Ray (Yuan, 1998). Tony award–winning choreographer and collaborator on "Beat It" Michael Peters designed the dance sequence for "Thriller." The choreography ranks among the most recreated dance sequences in the world and has become a staple of celebrations, marching bands, parades, holidays, commercials, flash mobs, and more. It also was integral to the success of the project due to its ability to harness the power of new media.

Prior to "Thriller," the majority of products that used new media replicated the traditions of old media. Its choreography created arresting visuals that propelled music videos to an art form similar to that of a Broadway spectacle. Subsequently, music videos were watched as well as listened to. This provided advertisers with captivated audiences and new opportunities to attain commercial success. Visual imagery for "Thriller" was further enhanced with the assistance of Hollywood makeup artist and special affects creator Rick Baker. The costume design was the result of the creative genius of film scholar and costume designer Deborah Landis. Landis had created iconic film costumes throughout Hollywood for such legendary film actors as Harrison Ford for his character Indiana Jones.

Jackson's iconic candy-apple red leather jacket with an M logo worn in the "Thriller" video has been referred to as the greatest piece of rock and roll memorabilia in history (Yuan). It became the most revered outerwear fashion trend of the 1980s and is widely emulated in virtually all circles even today. In 2011, nearly 30 years after the release of "Thriller," the celebrated jacket sold for $1.8 million in an online auction (CNN, 2011).

Contributing to the adoration and nostalgic sentiment surrounding the "Thriller" video was the addition of the sinister monologue, often referred to as a rap, by Vincent Price. It offered juxtaposition of old and new media that piqued the interests of multigenerational audiences. Price, a famed American actor well-known for his distinctive voice and performances in horror films during the 1950s, including the cult classics, *House of Wax* (1953) and *The Fly* (1958). Additionally, Price enthralled television audiences with his portrayal of a villain in the late 1960s *Batman* television series. In the 1970s Price also hosted and starred in BBC Radio's horror and mystery series *The Price of Fear* (Yuan). Price's noteworthy contribution in "Thriller" made the video synonymous with horror and the occult around the world.

The success of "Thriller" solidified a mass audience for the global economy. Moreover, it established a precedent for integrated marketing communications (IMC), branding, and the power of new media. Central to IMC is the practice of unifying all markets and communication, hence, "the big idea." Through multiple touch points and interactive experience, brand superiority was conveyed at a level that the world had never seen before. Synergy, created through consistency and coordinated communication, produced not only a superior product but also a memorable and emotional experience for audiences.

Connectivity was established through brand integration and lifestyle. Moreover, "Thriller" demonstrated that multiple stakeholders contribute to overall brand success. Everything was communicated through a unified team vision that showed both external and internal coordination. Bruce Swedien, musical engineer for "Thriller," recalls:

> That's why those albums, and especially "Thriller," are so incredible. The basic thing is, everybody who was involved gave 150 percent ... Quincy's like a director of a movie and I'm like a director of photography, and it's Quincy's job to cast. Quincy can find the people and he gives us the inspiration to do what we do (Lyle, 2007, p. 2).

With the success of "Thriller," Michael Jackson became one of the world's most beloved and celebrated icons. The brand promise of superiority had been communicated and delivered with unquestioned expertise. The natural progression of this IMC blueprint resided within brand relationships. Not only did brand relationships drive value, they also heavily influenced commodification.

Enterprises within American industry took notice of Michael Jackson's global success and innovation. Alan Pottasch, senior vice president for advertising at Pepsico, sought a brand relationship with Michael Jackson. Known as the father of the Pepsi Generation, Pottasch desired to invigorate the creative campaign and strategy he had devised in the 1960s and 1970s. Prior to Pottasch's arrival, Pepsi achieved success through product innovation. Accordingly, Pepsi was the first soft drink to introduce products such as two-liter sodas, twelve-pack cans, and environmentally friendly packaging. Pepsi employed entertainment marketing strategy, developing relationships with Hollywood icons and product placement with stars such as Joan Crawford. However, it was Pottasch who devised a creative strategy that would place Pepsi on par with its greatest competitor, Coca-Cola (Enrico & Kornbluth, 1986).

Under Pottasch's leadership, Pepsi launched the groundbreaking creative marketing campaign known as Pepsi Generation. This campaign focused on attributes of consumers rather than attributes of the product. Moreover, the central focus of this campaign was to connect with youth culture lifestyle within the baby boomer generation. This was an unorthodox and risky approach for any company during the 1960s. Pottasch affirmed:

> For us to claim and name a whole generation after our product was a rather courageous thing that we weren't sure would take off. Pepsi Generation became a part of the lexicon and pop culture. People referred to what we now call the Baby Boomers as the Pepsi Generation. That, of course, is a dream for a product. It made Pepsi part of everything that was going on, and that is a great place to be. (J. Stewart, 2007, p. 2)

Nearly two decades later, Pottasch sought to reinvigorate this approach, stating that he wanted to signal something new and nothing signaled new more than music. In addition to his role as senior vice president for advertising, Pottasch was Pepsi's primary coordinator with BBDO, one of the world's largest advertising agencies. In collaboration, they produced a new creative campaign, The Choice of a New Generation (McLellan, 2008).

During the time of collaboration, Philip Dusenberry was a creative leader at BBDO. Dusenberry ranked among the 100 most influential advertising figures of the twentieth century in a study conducted by *Advertising Age* magazine. Dusenberry set a precedent within the industry for raising advertising from an idea to an emotional and human experience. He was instrumental in creating memorable advertising campaigns for clients such as General Electric (We Bring Good Things to Life), Gillette (The Best a Man Can Get), Visa (It's Everywhere You Want to Be) and Home Box Office (It's not TV. It's HBO.). Dusenberry was a key member of the creative team behind an advertisement film for President Ronald Reagan's reelection campaign. Titled "A New Beginning," the short film introduced Reagan at the 1984 Republican convention. Dusenberry was also highly influential in the creative works for the nonprofit organization Partnership for a Drug-Free America (McLellan).

However, Dusenberry is best known for his work with Pepsi. Dusenberry was praised for the infusion of advertising into the realm of entertainment. His visionary creative productions rivaled Hollywood blockbusters and Las Vegas shows. In an interview with the *New York Times*, John Bergin, vice chairman of McCann Erickson Worldwide, proclaimed, "He [Dusenberry] does the sort of commercials that you call your wife and kids in from the other room to watch by saying: 'It's on again'" (McLellan, p. 1). Dusenberry won several distinguished awards, among them the Clio awards. He was also known for creating what BBDO termed as "advertising that touched the heart as well as spoke to the head." He affirmed, "I'm always going to be searching for emotion. In an age when most products aren't very different, the difference is often in the way people feel about the product" (McLellan, p. 2).

In 1984, the world was enamored with Michael Jackson. Not only had he been an American treasure for nearly 12 years prior, but now his explosive success demonstrated enormous promise that he would be a commodity for years to come. *Thriller*'s chart-topping triumph signified that he was the choice of a new generation and an ideal fit for Pepsi's campaign. Michael Jackson reworked his groundbreaking song "Billie Jean" and titled it "Pepsi Generation." It became the signature song for Pepsi's campaign launch party and was distributed as a promotional single. In turn, Pepsi was to sponsor Michael Jackson's Victory Tour, in which the Jackson Five was scheduled to appear. During the tour, Michael Jackson was to showcase his solo material from *Thriller*.

Moreover, the "Pepsi Generation" promotional single was used to create a jingle for the "Pepsi Generation" commercial in which Michael Jackson and his brothers would appear (Taraborrelli). Dance sequences from both the "Beat It" and "Thriller" music videos were demonstrated throughout the commercial. In an urban setting with the children of a new generation styled in iconic Michael Jackson fashions, Pepsi was self-proclaimed as the drink of the new generation. Pepsi was positioned as cool and hip through its association with the most celebrated pop culture icon of the time.

As a follow-up to the success of the first commercial, Pepsi intended to create another commercial that simulated the experience of a Michael Jackson concert. However, special effects malfunctioned. Pyrotechnics accidentally set Michael Jackson's hair on fire during the filming of the commercial. Jackson suffered second-degree burns and underwent medical treatment. After litigation, Pepsi settled out of court, and Jackson donated his $1.5 million settlement to the Brotman Medical Center in California, which now has the Michael Jackson Burn Center in honor of his donation (Taraborrelli).

Pepsi continued The Choice of a New Generation advertising campaign and expanded it with the inclusion of additional pop culture icons from multiple genres. Pepsi commercials starred such celebrities as Michael J. Fox, Madonna, and Lionel Richie. Through The Choice of *a New Generation*, Pepsi inferred that among soft drink consumers, there are Coke people and there are Pepsi people. If you were a Pepsi person, you were young, fresh, and cool. This campaign focused on portraying Pepsi drinkers as possessing desirable qualities, such as youth and esteem, through affiliation rather than on characteristics of the product itself. This creative strategy, along with the intermingling of music and iconic popular culture, was emulated throughout the industry (Enrico & Kornbluth).

Cultural Currency

Increasingly, advertising creative professionals became involved in the production of culture, as intermingling became prevalent within industry. Mergers fortified growing trends while also contributing to eventual massive domination. Media conglomerates supplied agencies with plentiful budgets to expand branding and products through lavish global campaigns and placement within music videos. Totaling upwards of $53 billion, advertising expenditures at national and international levels rivaled each other by the 1980s (Tungate).

In addition to Pepsi's The Choice of a New Generation, several iconic campaigns were extremely successful in selling products through lifestyle and cultural integration to a new generation of young, upwardly mobile consumers. Of particular importance were the campaigns for the signature brands Apple and Nike.

Through a fusion with pop culture, Apple also elevated the institution of advertising in 1984. In what has been hailed as a masterpiece, Apple introduced its personal computer to the American public. A commercial titled "1984" was broadcast on CBS January 22, 1984, during Super Bowl XVIII. Sir Ridley Scott directed the commercial. Scott, an English film director and producer, was popular for his commercial blockbuster films *Alien* (1979) and the science fiction classic *Blade Runner* (1982). The creative team of Steve Hayden, Brent Thomas, and Lee Chow at the advertising agency Chiat/Day conceived the commercial's concept. The commercial depicts themes of the literary classic, *Nineteen Eighty-Four*, a novel written by George Orwell in 1949. In a prevalent interpretation of the commercial, the female heroine represented the coming of Apple's Macintosh personal computer. The personal computer was associated with a means to save humanity from conformity, also known as Big Brother and Apple's primary competition IBM. In the novel *Nineteen Eighty-Four*, a dystopian future is ruled by Big Brother. However, advertising creatives insist that the commercial relied on borrowed themes from popular culture and resonated with audiences on multiple levels (Moritz, 2009).

There were numerous controversies surrounding Apple's "1984" commercial. Many considered the commercial to be a copyright infringement on the *Nineteen Eighty-Four* novel. However, Hayden insisted:

> The intention was to remove people's fears of technology at a time when owning your own computer made about as much sense as owning your own cruise missile. We wanted to democratize technology, telling people that the power was now literally in their hands. If you can remember back that far, the Cold War was still pretty hot. Reagan was in the White House, and the Soviet Union was the Evil Empire. We knew that if fax machines could bring down dictatorships, personal computers could do infinitely more. The Big Brother of the spot wasn't IBM—it was any government dedicated to keeping its populace in the dark. We knew that computers and communications could change all that. (2011, pp. 1–2)

Chiat/Day encountered many obstacles throughout the advertising commercial's production process. While on set, Apple's client representative refused to sign the estimate to permit the second day of shooting. Once the commercial was finally completed, the spot was very poorly received by Apple's board of directors. During an intense meeting a motion was made to remove Chiat/Day as Apple's advertising agency. Several Apple executives considered the commercial to be far too risky and did not initially support airing it during the Super Bowl (Hayden).

Nonetheless, the commercial did air. It was instantly hailed as a masterpiece within the field of advertising and is widely regarded as one of the most memorable and successful American television commercials of all time. Its fusion with popular culture contributed to its ability to be recognized as an artifact of nostalgia. In fact, "1984" was remastered and rebroadcast in 2004 for the twentieth anniversary celebration of the commercial. Moreover, "1984" established the precedent that the Super Bowl was a premiere platform that advertisers could leverage to launch innovative branding campaigns (Hayden).

Apple's "1984" was also significant because the commercial symbolized a shift in the advertising industry. Apple's "1984" was not created by one of the conglomerates in the industry. Although Chiat/Day agency was the result of a merger, it was not representative of a media powerhouse. The "1984" commercial helped Chiat/Day receive notoriety through the advertising industry for its creativity and production prowess. Chiat/Day was also located in Los Angeles rather than the traditional Madison Avenue or other East Coast locations.

However, after the success of Apple's "1984" commercial, the majority of the creative talent that worked on the Apple account was lured away to advertising conglomerations. This act also set a precedent that was subsequently emulated throughout the advertising industry. In 1986 Steve Hayden, a key creative in Apple's "1984," left Chiat/Day agency. Hayden continued his career at BBDO as the Chairman/CEO of West Coast Operations (Moritz).

In addition to influential precedents established through the works created by BBDO with Pepsi and Chiat/Day with Apple, Nike also offered a legendary contribution during the 1980s. Never before had a brand demonstrated such a keen understanding of consumer engagement with its product in the marketplace. The results of this awareness were phenomenal and had enormous cultural implications.

Nike itself began with a marked emphasis on creative ingenuity and strategic risks. Bill Bowerman, sports guru, acclaimed athletic coach, and Nike co-founder, was dissatisfied with available athletic shoe options for his athletes. Thus, he was inspired to create handcrafted running shoes for them and the groundwork for the impending brand. Originally known as Blue Ribbon Sports, Nike officially changed its name and launched a new campaign on May 30, 1978. However, it was not until 1985, with the signing of then Rookie of the Year Michael Jordan, that Nike became an icon of American culture and consumption. This period is referred to as the beginning of a golden run during which Nike was reinvented. It solidified a relationship between Nike executives and Michael Jordan in which they conspired to place their mark on general culture through their association (Strasser & Becklund, 1991).

In 1980 Nike decided to move their company forward in a broader direction. The company had primarily focused on running and providing athletic shoes for track and field athletes, runners, and joggers. Although running generated highly substantial revenue with sales totaling $236 million, there were new areas for growth and expansion. The overall athletic shoe industry indicated that the largest profits could be acquired through basketball and tennis endorsements. Nike decided to expand their product line in a variety of directions by adding performance features and moving into leisure and apparel. Nike also began to target new audiences, including nonathletes, and ventured into other sports, including soccer, tennis, and basketball (Strasser & Becklund).

Nike also made adjustments within its advertising strategies. In the past, the company had primarily focused its advertising budget on serious athletes. It had communicated with formidable athletes by using targeted sports media with messages about the technical superiority of the shoes, the integrity of competition, and the spirit of athletic endeavor. Nike executives believed that the use of celebrity athletes in its advertising should be minimized, under the assumption that such strategies demonstrated broad appeal to a savvy sports consumer adept at making choices. Nike executives were adamant that its consumers should not be made to feel manipulated through star emulation. Nike was built on the theory that advertising should appeal to elite athletes. Once elite athletes were won over, this success would trickle down to the masses. Elite athletes were considered to be those athletes who were authentic in their quest to achieve excellence in their sport. They were not primarily associated with celebrity or major league sports. Nike considered elite athletes to be students, coaches, and health enthusiasts.

Nike was wary that star emulation could create a disconnection among their core consumer audience. Nike wanted to remain focused on athletes and the spirit of performance, but the industry was moving in a broader direction. Aerobic and fashion trends moved athletic shoes into areas other than sports. However, Nike's competitive drive pushed the company forward and inspired change. Nike decided to nationally expand its brand and to continue building its image.

Despite enthusiasm, Nike found itself in an economic slump as it failed to leverage trends in the athletic shoe industry. Its chief competitor, Reebok, harnessed the popularity associated with fashion and aerobics and made significant strides in American domestic markets. In the 1984 Annual Report Nike executive Phil Knight affirmed:

> Our domestic footwear marketing is changing, edging away from athletic looks to a renewed demand for fashion and traditional styles. These changes resulted in inventory valuation losses over three times greater than in 1983.

Nike overlooked or dismissed a valuable trend within its market. Nike's creative guru Peter Moore suggested that Philip Knight was responsible for the oversight. He [Knight] missed, you know. He didn't miss often, but he missed that, clearly. Just didn't see it. Didn't believe that a woman's thing, one, could have that much impact; two, didn't believe that the soft leather thing could have any potential because it was not performance. (Coleman, 2010, p. 130)

Nike executive Charles Robinson argued that Nike's macho athletic ethos stunted the company psychologically and theoretically. Women were sometimes perceived in subordinate positions and therefore not taken seriously within the company or as athletes or consumers. As a result, Nike missed several lucrative opportunities (Strasser & Becklund).

In order to remain competitive, Nike made adjustments to its advertising strategies. In the past, it had primarily focused on promotions because it was believed they were more cost-effective than advertising. However, during these times, Nike was also plagued by the maturation and saturation of markets. More and more athletes entered the promotional market demanding increasingly large compensation for brand associations. Consequently, the value of promotional activities declined. As a potential remedy for its collective hardships, Nike decided to make cuts in athletic endorsements and focus its energies on superstars who could create significant influence. Additionally, Nike decided to develop advertising strategies based on a unique selling proposition (USP). Executives sought to emphasize the "air" technology as a product feature for their new advertising campaign.

Nike had long since determined the salience of the two-step flow of communication. In their efforts to connect with elite athletes, Nike established a lucrative practice of signing college basketball coaches. Nike's competition focused on openly buying professional athletes and sponsoring events, such as the Olympics. Nike forged relationships with consumers and potential superstars through the influence of athletic coaches. According to Nike executives, the company quietly acquired the best of the upcoming American athletes, "those gonna-bes that would give Nike the reputation of the shoe preferred by young, hot players" (Coleman, p. 126). This practice served as an impetus in the acquisition of Michael Jordan as a Nike icon.

The relationship between a company, brand, or product with a spokesperson had never before reached the depths or heights as that of the relationship between Michael Jordan and Nike. This became a partnership that was truly unprecedented and has henceforth been widely emulated. In essence, Michael Jordan fused with the Nike brand. This relationship superseded a mere celebrity endorsement. Both Nike's unique selling proposition and logo embodied Michael Jordon and became synonymous with his success and his superior athletic performance. During these times, embarking on such a significant relationship was risky. Nike had a great deal at stake but was fairly certain that they were making a huge investment in the future success of their company.

David Falk, Jordan's agent, negotiated an offer of $2.5 million over five years, plus annuities, signing bonuses, and most significant, a stake in the success of the Air Jordan signature brand (Strasser & Becklund). Instead of offering Jordan a lump sum only, Nike offered Jordan a percentage of the profits of the products bearing his name. This act symbolized Jordan's potential investment in the success of the brand and foretold of the impending brand personification. With a tremendous amount at stake for Nike, executives pushed for a collaboration of excellence surrounding this seminal creative endeavor. Advertising agencies were involved during the earliest inception of this undertaking. It was conclusive that the execution must be flawless. Nike executive Strasser stated:

> If Jordan does what we think he can, and if we can execute, this can be big. Nike is going back to sports, where we belong. These were a lot of ifs. The company was taking a huge risk. What if Jordan did not perform in the NBA? What if he was injured? Contracting Jordan would be a huge risk taken not only on his potential, but also on

the potential of the market to support signature products. He was little known by the public other than serious college basketball fans. (Coleman, p. 133)

Creative Director Peter Moore described the collaboration with Jordan as follows:

> It's pretty clear that he's a hell of a basketball player and he seems to have a pretty decent personality. At least our guys, the guys that were scouts for Nike, you know, say that the guy is a good kid. So Rob comes to me and says, I wanna get this guy because we need to get him, we need something to happen, but I don't just want to sign another basketball player. I wanna make something out, I think we can make something out of this guy. So, we meet with his agent, a guy named David Falk, and we go back and forth, and David Falk says I wanna call this thing Air Jordan. Um, and in addition to that, we built a whole program around it where it was an autographed shoe and nobody in the team business had ever done this. I mean, a signature shoe in a team business has not ever been a big thing. So we did it. (Coleman, p. 133)

Nike divided its creative agenda between multiple advertising agencies. The primary advertising agencies were Chiat/Day and Wieden & Kennedy. Executives and agencies began to conceptualize the creative image they would ask Jordan to build. Moore believed the name Air Jordan was an ideal fit for the company. In accordance with Nike's product features and unique selling attributes, Nike's Air technology facilitated an emphasis on innovative performance technology. Despite this innovation, the public did not quite grasp the technology (Strasser & Becklund). In an increasingly visually driven society, it became obvious that an icon was necessary for success. Hence, Michael Jordan became the clear choice.

> Nike had this thing called Air that nobody even understood yet, but we had it and it was sitting in these shoes. And this guy could fly and his name was Jordan, so it seemed a pretty good idea. (Coleman, p. 134)

The creative endeavors surrounding the Air Jordan campaign became known as the launch around the movement. Nike's intention was to shape culture and to create a seamless fusion between the brand and Michael Jordan. The company utilized integrated marketing tactics and a staggered campaign strategy to build momentum, interactivity, and create awareness. Up until this point, basketball players had traditionally worn white shoes. It was Nike's tradition to command attention through contrast. Nike creative director Moore acknowledged:

Nike had been one of the first companies to bring in bright colored basketball shoes. So, the idea was to say let's take this symbol, put him in some colored basketball shoes because he's going to be spectacular and let's call attention to him by doing this. Because otherwise, he is just gonna be another guy just wearing a logoed shoe. (Strasser & Becklund, p. 429)

It was integral to the success of the movement that distinctive parallels were drawn between Nike and Michael Jordan. Nike was known as the running shoe company that defied expectations by marketing bright bold athletic shoes. Because Jordan was quickly becoming known for his unprecedented talent, Nike executives believed that he also would be able to defy the rules. Hence, it was decided that they were going to break the color barrier on the professional basketball court. Moreover, Nike intended to build the campaign and movement around an autographed shoe, which had never been accomplished in team business. At Falk's urging, Nike treated Jordan like a tennis star and subsequently revolutionized professional basketball. In effect, Nike executives told Jordan that he was going to be a brand and that Air Jordan was going to be a subbrand of Nike (Lafeber, 2002).

Creative strategies and tactics fueled the Air Jordan advertising campaign. Nike's emphasis on color was used to draw attention to Jordan. Nike designed a signature shoe in the colors of Jordan's team, the Chicago Bulls. At the request of Nike executives, Jordan debuted his red and black Air Jordan Nike shoes in a preseason Chicago Bulls basketball game at Madison Square Garden in September 1984. Consequently, the National Basketball Association (NBA) banned the shoe for violation of the league's uniformity of uniform clause. The violation resulted in a $1,000 penalty for Jordan. Jordan was warned that the second violation would result in a $5,000 fine, and the third violation would result in a forfeiture for the Chicago Bulls.

Nike executive Strasser urged Jordan to wear the shoes for the opening home game. Strasser promised that Nike would cover the fine. Eventually, Nike pulled back due to the discontent expressed by Chicago Bulls general manager Rod Thorn. In compliance, Jordan wore a white Nike shoe with a red Swoosh for his first professional home game. However, in his next game, Jordan wore his new signature red and black Air Jordan shoes (Strasser & Becklund).

The *New York Times Magazine* reported that there was substantial backlash due to Jordan's refusal to comply with league rules and policies. Thorn, the general manager of the Bulls during this time, called David Falk, Jordan's agent, and complained, "Dammit, David, you're turning the guy into a tennis player" (P. Patton, 1986, p. 48). Although this statement symbolized negative backlash associated with the campaign, it also affirmed Nike's efforts. Turning Jordan into a tennis star was a strategic objective.

Attendance at the professional basketball games had risen significantly since Jordan's arrival and the Bulls were increasing their number of wins. Many contend that the exchange in the press was indeed choreographed to create drama and intrigue for the campaign and the league. Consequently, everyone involved—Michael Jordan, Nike, the Chicago Bulls, and the National Basketball League—benefited enormously from the publicity (Strasser & Becklund). Jordan's popularity quickly began to soar. As a result, Nike seized the opportunity to fuse the product, the demand, and the athlete to achieve an unprecedented level of success.

The next phase of the movement was a groundbreaking television commercial developed by Chiat/Day advertising agency. The commercial, titled "Jordan Flight," fused the undeniable grace and skill of Michael Jordan with Nike. The 30-second spot featured Michael Jordan simply bouncing a basketball, while the camera slowly panned him from head to toe. The voice-over said, "On September 15, Nike created a revolutionary new basketball shoe. On October 18, the NBA threw them out of the game. Fortunately, the NBA can't keep you from wearing them. Air Jordans from Nike" (Coleman, p. 138.).

The commercial relied on essential advertising tactics of exaggeration and puffery. "Jordan Flight" did not overtly state that the shoes were banned because their colors defied NBA rules. Rather, it implied that there was a connection between Jordan's ability and the revolutionary technology featured in the shoe. NBA Commissioner David Stern, who had banned the flashy shoes, eventually approved the advertisement. The benefits of Air Jordan outweighed any potential risks. During the times of the Air Jordan fervor, the NBA was experiencing its first revival since its last great superstar, Julius (Dr. J) Erving. Moreover, cable television ratings rose 20% in one year alone and game attendance was at a peak (Lafeber).

Although Nike secured approval from the league, there was still vast uncertainty overshadowing the movement. It was difficult to ascertain consumer response. Nonetheless, Nike aired Chiat/Day's commercial during the NCAA Final Four Tournament on April 1, 1985. The commercial supplied the public with an iconic visual for Jordan's developing trademark flight through the air. The commercial ended with a call to action, asking, "Who says man was not meant to fly?" Creative executives affirmed that the advertisement was a testimonial of Jordan's ability to fly like a bird and the implication was that the padded technologies bound to his feet had something to do with his agility and grace. The Air Jordan line was released in stores later that month. The initial sales projections of 100,000 pairs of Air Jordan shoes were exceeded by more than 1 million. Sales soared well over $100 million in Jordan's first season of professional ball, 1984–1985 (Lafeber). Steve Aschberner, sports writer for the *Chicago Journal*, affirmed the popularity with his public praise: "Michael Jordan is not the most incredible, the most colorful, the most amazing, the most flashy, or the most mind-boggling thing in the NBA. His shoes are" (Strasser & Becklund, p. 451).

Creative direction capitalized on publicity to increase the momentum surrounding the appeal of the shoes. Creative director Moore referenced the creative strategy as follows:

> So, anyway, we ended up signing him. *Big* deal, got the shoes made, then had the shoes banned which was even better because now everybody wanted to buy them. And we did the ad, we did a television thing around him and I think the closing line was, who said man was not meant to fly? (Coleman, p. 138)

Nike entered the cultural fabric of American society. One retailer was quoted as saying, "Imagine, these young kids are coming in who will grow out of them in a few months and their parents are buying them at $64 a pair" (Lafeber). The sneakers had become more than simply shoes. Like Jordan, the shoes were a symbol of greatness. Jordan was portrayed as a mythological hero who could fly. Through the use of esteem and affiliation, the shoes became a symbol of status and stature. The public demand for the shoes was astounding and resulted in even more controversies. The public demand for Air Jordans created a situation in which both consumers and retailers inflated the price of these coveted shoes to sell them for their own personal profit. NBA spokesperson Terry Lyons was asked if he had a pair of Air Jordans. Lyons retorted, "I can't wear them around here, I'm afraid they'd be taken away from me on the way to work by some kid with a gun" (Coleman, p. 141). This statement not only affirms the demand for the Air Jordan shoe but also symbolizes the cultural obsessions and controversies associated with this brand.

Nike creative director Peter Moore stated the importance of creativity and strategy throughout the movement as follows:

> That television commercial, together with the shoes, together with the poster of the flying man, together with the whole in-store thing, it was just an instant success. It was just bam. And that showed us that with a small group of people—the whole

thing was put together by 8 people at Nike—controlling the whole package from the product to the advertising to the in-store to the way the kid dressed and acted, you had this whole entire thing together and you control the image. And you could also control the volume. How loud did you wanna make it? How scarce did you wanna have it? We had kids, we had people lining up in stores before they opened, an hour before they opened, to get this product. You know, it was a phenomenal thing. And it was the first time it had ever been done. And it did break the fever. I mean, it did take, it broke the fever, didn't solve the illness, but it broke the fever. (Coleman, p. 141)

With the astonishing success of Air Jordans, Nike decided to tighten its creative team in efforts to sustain success and brand integrity. The company began an exclusive partnership with Wieden & Kennedy, despite the success achieved with Chiat/Day. Nike released the following statement affirming its new partnership objectives:

We want to continue to work with a tight group of people. We are not interested in planning, research, or squads of account people. We want chemistry and creative force, not account service or media buys. I want mistakes, I want f***** ideas. Failure makes for better, in a lot of cases. I want people who listen, who hear, who think, who bleed, who laugh, who don't worry about time sheets and looking good. People who sweat, who work and who *love* what they are doing. (Coleman, p. 128)

Nike and Wieden & Kennedy began to cultivate a fruitful relationship that earned acclaim for both the advertising agency and the brand, eventually rising to the top of their respective industries. As Dan Wieden articulated, "A lot of our creative outlook ... and frankly some of our management style has been developed by working closely with Nike" (Coleman, p. 124). They established a mutually beneficial workflow that was highly influential within both companies.

The partnership forged between Nike and its creative agencies demonstrates the significance of synergistic relationships. It is also symbolic of the level of success that can be achieved through collaboration and risk taking. Rob Strasser, Nike executive, expressed what a Nike advertising agency should be:

Authentic. We know sports and we think we know guys. The agency must know sports. They don't have to be stat freaks or trivia buffs, they have to know sports. Competitiveness, camaraderie, face, all the gut that makes sports, sports. They must feel or know the thread that weaves its way through the athletes. That same thread in Moses Malone, Alberto Salazar, John McEnroe. They must know guys. (Strasser & Becklund, p. 458)

The partnership between Wieden & Kennedy and Nike continued to fortify the Nike empire. Together, they produced riveting creative campaigns and branding that continued to defy expectations, set trends, and push the brand to the next level. Again, they made history and fused even further into American culture with their Spike and Mike commercials. These commercials costarred acclaimed director and actor Spike Lee. In the commercials Lee appeared as Mars Blackmon, a character he had first portrayed in his 1986 breakthrough film, *She's Gotta Have It*. These ads depicted an impressive knowledge of cultural trends and the latest in advances in new music and film techniques. This campaign also contributed to the elevation of advertising and its fusion within popular culture. It shifted general attitudes surrounding sports to levels that were inspiring and revered by the public. The Bo Knows campaign for cross-trainer athletic shoes followed the Spike and Mike campaign. The campaign featured the latest computer graphics technology and sports megastar Bo Jackson.

Even further success was achieved with the carefully crafted Charles Barkley I Am Not a Role Model campaign, as described below:

> Nike Air "Revolution" campaign, the Spike Lee/Michael Jordan commercials and the "Just Do It" campaign series with "Bo Knows" ... these campaigns became deeply engrained in American culture and vernacular, and secured Michael Jordan's and Bo Jackson's places as the first and second most famous athletes in the world and placed them on the thrones of demigods ... Nike, as any company, continued to rise and fall in the market, but these campaigns helped to make Nike an icon,—the swoosh, the revolution, the empowered call to action "Just Do It," the company of Michael Jordan, the once counterculture running shoe company that became a badge and a casualty of urban cool. (Coleman, p. 161)

Campaigns such as these have contributed to advertising culture in America and abroad. Through a fusion with creativity and culture, advertisements have become exalted as artifacts of popular culture. Commercials are now an intrinsic aspect of modern society. They often transcend the space contained by their literal meanings. They are iconic representations of culture, which reflect social trends, aspirations, and values. In some instances, commercials and advertising campaigns have become the prized possessions of modern civilization. This symbolism foreshadows even deeper intermingling between advertising and culture as the American public has demonstrated its reverence and obsession for brands.

The phenomenal success of Nike helped to create a powerful model for advertising. Subsequently, the industry witnessed a pervasive emulation of hip-hop and urban culture. Advertising hoped to recreate Nike's enormous success and brand reverence. Similar tactics were used to integrate brands in an effort to reach youth culture in an equally lucrative capacity.

By the late 1980s the American public, youth culture in particular, had grown tired of advertising and had grown weary of its tactics. An *Advertising Age* study conducted in the early 1990s revealed that Americans considered advertising less during purchasing decisions than in previous years. Of the participants in the study, fewer than 15% relied on advertising in buying appliances, 10% or fewer in buying furniture, 7% or fewer in making banking decisions, 9% or fewer in buying automotive supplies, and 17% or fewer in purchasing clothing.

Further research revealed that generation X, which at that time was comprised of 47 million 17–28-year-old Americans, was a generation lost to advertising. Generation X was virtually immune to the strategies and tactics of the advertising industry. Sandwiched between 80 million baby boomers and 78 million millennials, generation X has been roughly defined as anyone born between 1965 and 1980. Generation X has a mere 46 million members, is highly educated, technologically savvy, racially and ethnically diverse, and known for its unique and independent thinking (Gordinier, 2008). Such a small, yet diverse, demographic can pose a problem for advertisers. Advertising struggled to reach this audience. Traditional advertising strategies and tactics failed to counteract the new formulated norms surrounding this particular youth culture. Generation X was not receptive to traditional advertising. It became apparent that the product was no longer the hero, and lifestyle and culture integration were the newly pursued directions.

· 5 ·
PARADOX & PITFALLS

The 1990s were filled with seemingly advantageous opportunities for advertising. Advances in technology provided superior workflows, enhanced equipment, and new ways to reach mass audiences. Despite the promise associated with these opportunities, advertising began to struggle. Its drift toward decline quickly spiraled into relentless chaos.

Advances in electronification led to democratized information and technology within advertising. As a result, advertising experienced major disruptions within its internal structure and consumer environment. Disruptions were largely attributed to technological contributions from Adobe and the Internet. These disruptions eventually led to the cessation of productivity and power redistribution, which signaled the impending paradigm shift.

Adobe's technological contribution was epitomized through its release of the computer software applications Illustrator and Photoshop. This technology revolutionized the internal structure of the advertising industry in numerous areas, including workflow, skills, and preparation. Significantly, this technology diminished the strongholds of conglomerations, clients, and vendor relationships. Advances also led to increased accessibility, improved equipment, and lower costs. Accordingly, advertising professionals and creative practitioners began independent ventures, which led to increased competition and factions within advertising.

Moreover, the rise of the Internet in the early 1990s created a global mass communications phenomenon. The Internet quickly became an indispensable element to lifestyle and professional activity. The Internet facilitated the flow of information and technology while connecting users worldwide. It influenced generations, communication practices, and developed distinct behaviors and subcultures.

While the Internet offered invaluable contributions, it posed numerous challenges for advertisers and advertising agencies. The Internet provided consumers with information and exponential options, which led to migratory behavior, interactivity, and comingling. These challenges disrupted the consumer environment as it became increasingly difficult to influence modern consumers. Moreover, the inability to successfully monetize the Internet plagued advertising.

Declining conditions were exacerbated by the rapid progression of technology and unpredictable consumer behavior. By the close of the 1990s, advertising was enveloped in a crisis. This period of instability was compounded by even more hostile and radical change. By the 2000s advertising was fragmented and suffered the repercussions of unfavorable partnerships. Boutiques and smaller independent agencies flourished in unprecedented locations including Minneapolis, Minnesota; Portland, Oregon; Richmond, Virginia; and Peoria, Illinois. There was a major shift away from traditional Madison Avenue advertising. Moreover, the dot-com bust left advertising reeling from devastation. Unlike the aftermath of previous economic downfalls, advertising was already suffering a state of crisis. The devastation experienced from even further loss worsened conditions.

By the time social media became popular around 2005, advertising had greatly deteriorated. Collective productivity came to a thundering halt. Advertising suffered from a litany of issues. The most pressing concerns included factions, monetization, the rapid progression of technology, massive layoffs, professional restructuring, erratic consumer behavior, the uncertain future of adverting agencies, the preparation of future advertising professionals, and disintegrated relationships with clients and vendors.

The exact delineations of the crisis in advertising are difficult to extricate. Increased blurring intensified advertisings interdisciplinary nature and led to further complexities. Some branches of advertising suffered for years as their specific interests went unrecognized by the majority of the profession. Conversely, there were branches of advertising that achieved success. However, the productivity of the discipline reached a standstill. Within the explication of crisis insights emerge that will assist in the ability to navigate through complexity.

Democratized Information

By the close of the 1980s, there was yet another powerhouse that would have a tremendous impact on advertising. In 1982 Adobe was a fledgling company. Democratization of information had been brewing since its inception. By the close of the decade, Adobe had grown to provide a vital platform for consumers. Compounding this fact, Adobe would revolutionize the business and commerce practices of nearly every major creative industry.

John Warnock and Charles Geschke invented Interpress, a page description language that would facilitate a seamless translation of information from screen to press. This achievement became the cornerstone for the formation of Adobe in 1982. In 1984 Adobe released its first product, PostScript, a computer software language that was device independent. Moreover, its syntax was freely available. PostScript quickly became the language of the desktop publishing world and would eventually serve as an industry standard and universal language. PostScript provided a practical alternative to the restrictions and complexities of the print publishing industry (Adobe, 2007).

In 1984 Adobe partnered with Apple to place its PostScript language on Apples's latest computer systems. Apple was slated to introduce its new personal computer to the world in 1984. Adobe's partnership with Apple drove PostScript language into the mainstream market. Adobe followed its technological success with additional advances and Adobe and Apple expanded their partnership. Adobe created a PostScript controller for Apple's laser writer printer (Adobe). Consequently, the public possessed industry standard printing capabilities at a remarkably lower price. Additionally, Adobe made strides with the release of its typeface software technology. The company soon became established as a technological innovator that produced reliable and high fidelity products and solutions.

Adobe realized that the equity in its brand and products relied heavily upon their application. Consequently, the next major objective of the company became the creation of premiere applications to optimize PostScript language. Adobe reached an apex through its development of computer software for creative professionals. In 1987 Adobe introduced Illustrator, a vector-based drawing program. Adobe Illustrator quickly became an industry standard. Adobe Illustrator facilitated the creation of fonts, shapes, and Bezier curves with unprecedented accuracy. The mastery of Bezier curves, mathematically based equations used in image manipulation computer software, made Adobe Illustrator resolution independent. This largely contributed to Adobe Illustrator becoming the ideal application for the creation of logos and branding solutions. Two years later Adobe introduced Photoshop, which became its flagship product. Adobe Photoshop is photo and graphics manipulation and editing software. Adobe Photoshop soon dominated production and design in virtually all creative industries (Adobe).

As the years progressed, Adobe continued in the vein of seminal technological advancement. Adobe heavily contributed to the democratization of information. This had a profound impact upon advertising, as well as its business and practices. Technological achievements produced by Adobe resulted in both significant progress and devastating setbacks. Professionals, hobbyists, and novices alike now had access to industry-standard production. As a result, advertising witnessed an influx of unskilled practitioners who were versed in Adobe software but were neither firmly rooted within the disciple nor recipients of adequate training and education. Moreover, colleges and universities struggled to keep pace with the changes in the industry, which resulted in shortcomings in curriculum and inadequate preparation of students entering into advertising. Practitioners currently within the field had to grasp the new technology. Constant upheaval in advertising led to constantly changing job descriptions and responsibilities. Some practitioners quickly adjusted, while others struggled with change. Moreover, vendor relationships suffered. With the production industry experiencing significant transition, many advertising agencies began to outsource or develop in-house production capabilities (Adobe). Thus, the integrity of the discipline suffered.

Numerous critics of Adobe software argue that it contributed to a decline in ethics within advertising. In addition to industry woes, it has become virtually impossible for viewers to differentiate between manipulated imagery and reality, which significantly impacts the credibility of professional practitioners. Nonetheless, supporters contend that Adobe technological software advances have had positive implications for artist creativity and have produced superior workflows. Moreover, with increasingly reliable technology at substantially lower costs, advertising professionals have been able to venture out to create companies of their own and to increase their consultation and freelancing opportunities.

As agencies expanded during the 1980s, stifled creativity and tension often followed. In many cases, larger agencies could not freely provide creative flexibility and specialized attention to smaller clients due to budgetary and research restrictions. Small and independent agencies were established as targets during a widespread trend of mergers and takeovers.

In 1987 in the industry's first hostile takeover, the WPP Group purchased J. Walter Thompson (JWT) for $566 million. Two years later they acquired the Ogilvy group for $864 million. The holding company, Omnicom, was formed in 1987, which combined DDB and BBDO, among other notable agencies ("Advertising: A History from 1980–Current Day," 2009). Thus, factions and discontent were produced within many agencies. Technologies offered by Adobe contributed to a renewal of entrepreneurial spirit in the industry while also furthering an increased instability. Although the advertising industry had become accustomed to double-digit growth, from 1985 to 1990 advertising industry growth fell nearly 8% (Tungate, 2007). Previously, advertising agencies were prospering and could replace talent or provide incentives for employees. However, in these times of economic hardship, advertising agencies were not only devastated by a loss of creativity and talent but also undercut by competition.

As the decade progressed, several factors characterized an impending crisis, the most pertinent being media fragmentation. Media choices proliferated with new technology, which resulted in increased consumer-audience fragmentation (Jenkins, 2006). Advertising's traditional strategies and media selections were no longer sufficient to command the evolving consumer market (Rust & Oliver, 1994).

Accordingly, advertisers widened their sphere of influence and range of practices in order to remain relevant and influential. The partnership between popular culture and advertising was further solidified through renewed brand integration and product placement. Advertising began to undertake an increased role in the production of popular culture in the United States. The advertising industry vehemently embraced sales promotion, direct marketing, and home-shopping techniques, which further altered the discipline and created reverberating shifts (Jenkins). American cultural historian T. D. Taylor states:

> This shift is part of a series of vast changes in the realm of cultural production, changes so all-encompassing that it is now possible to argue that we are living in a moment when traditional cultural hierarchies are in a greater degree of flux than they have been in quite some time, perhaps not since the great upheavals of the nineteenth century that were cogently described by Lawrence W. Levine in *Highbrow/Lowbrow* (Levine 1988). The instability of today's situation is the result of a number of factors:

the ever-growing role of advertising, which continues to identify and capitalize on the trendy, sometimes even becoming the arbiter of the trendy; the convergence of content and commerce in cultural production; and the rise in eclecticism in the taste of social elites. (Taylor, 2009, p. 406)

Concurrently, the deregulation of much of American enterprise greatly affected the advertising industry. This deregulation resulted in increased partnerships, disciplinary sects, and a growth of widespread influence (de Sola Pool, 1983). Moreover, banks and corporations grew enormously in accordance with deregulation policies and rulings by the Federal Trade Commission (FTC) and the Supreme Court. These shifts ushered in new patterns of cross-media ownership. Previous barriers separating media began to disintegrate as digitization facilitated the conditions for the crisis (Jenkins).

Triadic Convergence & Radical Change

De Sola Pool asserts that prior to shifts spurred by convergence, media had its own distinct functions, markets, regulations, phones for conversation, radio for news, television for entertainment, movies for drama, compact disks for music, print for text, and so on. These distinctions were largely the byproduct of political choice and business practice rather than driven by the essential characteristics of media or technology (Jenkins). Pool suggests that central control is more likely when the means of communication are concentrated, monopolized, and scarce. Conversely, freedom is fostered when the means of communication are dispersed, decentralized, and easily available (Jenkins).

It was during these originating times of media conglomeration that the term "convergence" was first introduced into mass communications' scholarly discourse. In regards to mass communications, convergence first appeared in Ithiel de Sola Pool's *Technologies of Freedom*:

> A process called the convergence of modes is blurring the lines between media, even between point-to-point communications, such as the post, telephone and telegraph, and mass communications, such as the press, radio, and television. A single physical means—be it wires, cables or airwaves, may carry services that in the past were provided in separate ways. Conversely, a service that was provided in the past by any one medium, be it broadcasting, the press, or telephony—can now be provided in several different physical ways. So the one-to-one relationship that used to exist between a medium and its use is eroding. (p. 23)

A period of flux within advertising was met with even more hostile and radical change. The rise of the Internet in the early 1990s created a global mass communications phenomenon. Although the Internet had been around for decades, the commercialization of networks that occurred in the 1990s resulted in its exponential popularity. The Internet was quickly approaching billions of users worldwide and had become integrally fused within the operations of government, institutions, commerce, and industry. It also was becoming a central and indispensable element at the core of lifestyle and activity, influencing generations, communication practices, and developing distinct behaviors and subcultures.

Like nearly every aspect of society, advertising was deeply affected by the rise of this new and groundbreaking technology. The Internet altered virtually every facet of advertising from its agency environment, strategies, and practices to its markets, consumer audiences, and underlying business principles. Moreover, the Internet was reshaping and redefining media. In its early stages, the most prevalent concern was the transition of advertising to this new media. Ergo, Internet advertising debuted in 1994, with the launch of *Hotwired*, a commercial web magazine. *Hotwired* charged sponsors a fee of $30,000 to place an advertisement on its website for 12 weeks. Initial sponsors included AT&T, MCI Communications, Club Med, Adolph Coors Co.'s Zima brand, IBM, Harman International Industries' JBL speakers, and Volvo Cars of North America ("Ad Age Encyclopedia of Advertising: E-commerce," 2003).

In actuality, the transition to a new media was just one of a plethora of challenges faced by advertising. One such challenge was audience fragmentation, an already plaguing issue within advertising, that was now further exacerbated by the rise of the Internet. Consumers were provided with more choices, more control, and a greater capacity for interacting with sources of information. Accordingly, the mass media environment grew increasingly sophisticated and expensive. In the past, advertising shifted its most lucrative media practices and principles from old media to new media. However, the Internet required dynamic understanding. Advertising struggled as instabilities converged.

In regards to agency environment, significant displacements occurred as a result of the Internet. Due to the complexity of this new media, clients often contracted consultants in such areas as database marketing and interactive media with the purpose of meeting diverse racial, ethnic, and value-driven requirements. Specialized consultants were inserted into the traditional relationship between agency and client. Advertisers were discontent with the services and treatment received from large agency-holding companies. Consequently, larger agencies began to be perceived as aloof and unresponsive. To safeguard against these allegations, agencies began to expand their services and capabilities to an even greater extent.

Despite the best attempts, new media was itself in flux. Developing an understanding to aid predictability and control of the Internet was fruitless; therefore, advertising entered a period of intense transition. The size, structure, and functions of agencies began to change, along with the nature of advertiser-agency relationships. Some clients began to consolidate accounts at fewer agencies, while others acquired services from several agencies or dissolved partnerships altogether. Still others decentralized aspects of their decision making, allowing brand, category, and regional managers to make decisions on advertising and promotional activities as dictated by market forces. Additionally, desires for global reach and unsaturated markets began to place a heavy burden on American agencies. Budgets were slashed as advertisers began to cut back on advertising spending. Both clients and agencies were under tremendous pressure to discover the most effective media outlets and reach the largest audience at the lowest possible cost.

In addition to agency tension, the impact of convergence was also evident in advertising's creative works. Advertising campaigns and techniques incorporated media, technology, and culture. In a significant departure from television, which placed primary focus on visual imagery, the Internet placed a renewed emphasis on information and copywriting. Language is comprised of powerful symbols that reflect social trends, identity, and culture. Consequently, copywriting plays an instrumental role in advertising and popular culture. The use of slang, phrases, and patterns of mannerisms create lucrative passion points among audiences.

Accordingly, the proliferation of short message language systems greatly influenced culture during the 1990s. This comingling was leveraged by the advertising industry in the 1990s, resulting in a new form of copywriting that coincided with the explosion of the Internet. Shifts in advertising campaign–writing techniques correlated with advances in mobile devices, technology, and interactive platforms. Subversive spelling, brevity, phonetic language, hip-hop style, Internet shorthand, and cultural terminology reflected major trends in convergence demonstrated within advertising creative.

By the early 2000s automobile companies demonstrated this trend. Campaigns were developed to appeal to new generations of consumers who were accustomed to technologically driven and culturally relevant communication styles. In 2001 Cadillac unveiled its new car model driven by the IMC campaign for CTS. Acura soon followed suit with its line of models TL, RL, MDX, and RSX. Budweiser sought to remain relevant through a similar technique with its highly successful campaign Whassup! This phrase, taken from hip-hop and urban culture, was reflected in a short underground film written by Charles Tone. This film caught the attention of Budweiser advertising executives and was leveraged to create a mainstream sensation. The campaign garnered numerous prestigious awards and became a revered trademark of Budweiser. Boost Mobile, a cell phone company, employed a similar copy-driven technique with Where You At? This advertising campaign emphasized the walkie-talkie and GPS application features available on their mobile devices. Boost Mobile's marketing originally sought to connect with sports-driven enthusiasts, but it soon expanded its audience and emphasis to address urban lifestyle (Danesi, 2012).

In addition to influencing advertising creative copywriting techniques, advances in technology caused significant shifts in other areas of advertising. Internet technologies made the physical location of an agency virtually irrelevant. Through the Internet, advertisers and advertising agencies were provided new forms of communication to receive and provide services. There were also formidable technological advances in computer hardware and software used to create and alter images. Commercial spots and shoots that once required costly budgets due to location and scouting could now be produced through the use of less expensive and convenient computer graphics. As a result, trends of artistry and computer graphics were prevalent in the 1990s. Coca-Cola achieved success using computer graphic techniques with its Polar Bears campaign ("Advertising: A History," 2009).

Advances in technology, including the Internet and computer software, not only produced distinctive graphic techniques but also empowered small creative boutiques. Successful regional shops sprang up in fresh locations such as Minneapolis, Portland, Richmond, and Peoria. This contributed to a decline in creative talent within large advertising agencies and increased competition within the industry. Well-established agencies often counteracted, creating small niche units to serve clients in specialized areas. Concurrently, agency-holding conglomerates created media departments. These departments provided full media services to clients. They also gave the appearance of a different media agency but were actually a division within an agency-holding conglomerate. Nonetheless, these media departments permitted agencies to seek clients outside of the scope of their parent relationships and offered clients increased options. Leading independent media companies that emerged from traditional agencies in the 1990s included Mindshare from WPP Group, Worldwide from Omnicom Group, Zenith Media, and Initiative Media Worldwide from Interpublic Group ("Advertising: A History").

In addition to spurring changes that influenced agency structure, the Internet spawned a myriad of advertising strategies and practices. There were several factors that compounded the complexities of the Internet, including content and commerce, which were more closely integrated on the Internet than in other forms of advertising. Additionally, rapid technological advances created a steady stream of change, along with facilitating convenience and power to users. Last, the majority of Internet content is free, which made monetization a topic of astronomical importance.

Despite challenges, the Internet offered the promise of a global mass media. Moore's law (1965) indicates that the number of transistors on integrated circuits doubles approximately every two years. Moreover, Metcalfe's Law (1980) states that the value of a telecommunications network is proportional to the number of connected users of the system (Hendler & Golbeck, 2008). Hence, the strength and value of interconnected networks linked by the Internet grows in exponential proportion with each new user. Radical and unprecedented interaction solidified common interests and joined numerous entities of diverse backgrounds. Moreover, digitization made it desirable and lucrative for companies to distribute content across channels rather than a single media platform. Digital scholar Nicholas Negroponte coined the term "digitization" as a way to describe the transformation of atoms into bytes (Negroponte, 1995). Such transformative technology fueled the desire of droves of investors who sought to harness the explosive power of the Internet.

Associations & Accountability

History has indicated that nearly every great wave of transformative innovation is accompanied by financial mania. The Internet was no exception. Canals, the telegraph, railway systems, and automobiles have all been surrounded by similar patterns, also known as economic bubbles. History suggests that the cycle is constant with breakthrough technology, countless risky startup businesses, and the promise of fortune, which is typically followed by droves of enthusiastic investors. History also suggests that economic bubbles eventually burst in the form of colossal bankruptcy. The competition created by excess capacity inevitably leads to vicious price competition and a failure to sustain profit. Enormous economic upheaval is unleashed.

The pernicious downsides of economic bubbles are devastating. The financial losses are tumultuous and plainly evident. However, the repercussions within interconnected industries, lifestyles, and society are often less obvious, compounded, and resound for many years afterwards (Gross, 2007). Advertising was devastated by losses associated with the economic bubble surrounding the Internet. Advertising's intricate relationship with business and the American economy intensified instability spurred by convergence. This led to deepened levels of crisis.

Despite potential risks, numerous advertisers were eager to forge alliances to partake in venture capital cash endeavors (Jenkins). As a result, advertising agencies were often thrown into the center of mayhem. Although the dot-com frenzy brought a great deal of desirable revenue into advertising, it heightened already tense client relationships within the industry. A number of advertising agencies found it difficult to work with dot-com clients. From the outset, any new company seeking to utilize the Internet as an advertising medium faced serious challenges. Unfamiliarity and skepticism shrouding this new media hindered its credibility. Traditional advertising agencies approached new projects with caution. Agencies believed that client relationships, as well as strategy, would be integral to the success of any works created. Many of the dot-com entrepreneurs sought traditional ad agencies to assist with building brands and creating awareness. However, it became evident that there was a major culture clash, which inhibited progress.

Decades of advertising created traditions and protocols that placed a high value on long-term, agency-client partnerships and structured methods of evaluating and reviewing strategy. These practices were the result of years of business with established marketers that were risk averse and preferred methodical decision making. On the contrary, dot-com companies tended to be driven by a spirit of pioneering innovation and risky decisions. Dot-com entrepreneurs were likely to be averse to precedents and restrictions. Consequently, several top agencies refused the business of dot-com clients. This created a trend within the industry. Goodby, Silverstein & Partners became suspicious of their relationship with PlanetRx.com, an online drug retailer, after merely a few months. Goodby, Silverstein & Partners claimed that PlanetRx.com demanded an immediate campaign but could not make basic decisions on goals. Goodby, Silverstein & Partners ended the relationship when PlanetRx.com brought in another agency as partner to undertake a separate project (Inukonda & Pereira, 2010).

In contrast, scores of advertising professionals were both excited and intrigued by dot-com start-up companies. Several chose to be vanguards within this emergent era and left agency positions to do so. This led to further dissention within the discipline and factions that rose throughout the field. Critics argued that advertisers flocked to this new media technology without full comprehension of its inner workings. Others argued that advertising was inadvertently caught in an inextricable web of business methods and practices. Nonetheless, shallow solutions were generated to produce hyper-rapid awareness but lacked formidable advertising and branding techniques. In 1998 the top 50 American Internet advertisers spent $420 million on advertising. By the following year, spending had increased more than 280% in two months alone. Moreover, similar situations were occurring simultaneously worldwide (Tungate). Thus, the dot-com frenzy led to major contention within the advertising discipline. Experts warned that successful brand building took place over time. They cautioned agencies not to yield to the promises of fast money and fame. An article in *New Age Media* titled "How Dotcoms Killed off the Ad Agencies," characterized the period as follows:

> Bewildered middle-aged agency suits sat open-mouth as teams of idiots in three quarter length trousers and Japanese trainers hosed backers' money at them ... They took briefs from 26-year old marketing directors whose sole previous experience had been the production of club fliers. (Tungate, p. 214)

Nonetheless, the dot-com boom continued well into the 1990s, with multi-million dollar accounts, advertising campaigns, and television commercials. From February 1998 to October 2000, 413 dot-com marketers spent an estimated $6.5 billion in advertising; ironically, much of it was spent via old media vehicles like television. With the market now flooded with e-brands, clients were simply fighting for name recognition. The dot-com marketing frenzy peaked with the airing of Super Bowl XXXIV in January 2000. Seventeen dot-com companies advertised and accounted for nearly half of all the event marketers. The dot-com companies paid an average $2.2 million for a 30-second commercial spot (Inukonda & Pereira).

By 2000 the dot-com frenzy ended in grisly demise. The NASDAQ market had reported a loss of more than US $3,000 billion in value since is previous peak (Tungate, p. 216). It was declared that the year 2000 would go down in history as the year that more stock market wealth was destroyed than ever before (Tungate). Within the advertising industry, online spending froze. The Internet Advertising Bureau confirmed that, after soaring to rates of 150%, spending in one quarter tumbled to a negative 6.5% by the close of the year (Hovland & Wolburg, 2010).

At the close of the catastrophe, the Internet remained. The Internet was still relatively new and a largely untapped, powerful medium. As an upside to the devastation, the dot-com bubble created a commercial infrastructure that sustained global networks. Fiber optic channels, ethanol plants, and web-hosting facilities had been developed at a rapid pace to coincide with dot-com fervor. In addition, successful Internet companies that survived the bust, namely Google, Amazon, and eBay, provided valuable models for the future (Simon & Joel, 2011). The frenzied pace of Internet development associated with the dot-com boom also helped to alleviate skepticism associated with the Internet. The dot-com boom fostered curiosity, comfort, and intrigue surrounding the Internet. It allowed people, businesses, and markets to engage in new ways. Consequently, the Internet expanded to become a global center for commerce, education, entertainment, and social exchange.

Although there were many benefits, there were also substantial setbacks. Advertising was left reeling from the devastation associated with the dot-com bust. Unlike the aftermath of previous economic downfalls, advertising was already suffering from endemic turmoil and factions. The devastation suffered by the dot-com bust worsened schisms within the discipline. Moreover, advertising was bewildered in a new media environment with unprecedented diversity and increased levels of interactivity. Disunited, with few certainties and very little understanding, advertising split in multiple directions.

A survey of the advertising landscape post dot-com boom revealed extensive dissension. Although some aspects of advertising continued the slow movement forward, a great number did not. There was no unified paradigm to create momentum to move the discipline, in consensus, forward. Problem solving was limited. It became evident that interdisciplinary efforts and the strategic use of modernity could no longer sustain advertising. Psychology, business, and art had reached inharmonious accord. Psychology had shifted and long since moved in a direction away from behaviorism. As a discipline, psychology now placed a larger emphasis on cognition, self-mediation, and human agency. Business itself was in a state of crisis and flux due to global economic turmoil, saturation, maturation, and fragmentation of markets. Fine arts struggled to determine its place within advertising. Moreover, the creative world was undergoing its own metamorphosis due to advances in technology and computer software ("Advertising: A History," 2009).

Advertising's strategic use of modernity became virtually inconsequential when compared with the devouring pace of contemporary technology. This rapid growth far surpassed what could be decisively generated through demand creation. Branding became increasingly taxing. Markets were flooded with parity products and user-generated content. Moreover, the technology and media that often sustained and rejuvenated advertising in times of turmoil was an enigma and plague to its structure and foundational core. Geodemographic clustering technology was grossly insufficient. There was no longer a general mass market. Advertising could not easily dictate the needs and wants of all consumers. Decades of interactivity had coalesced with culture and lifestyle, amalgamating needs, wants, desires, and interests.

With record levels of internal and external strife, advertising ventured in a number of different directions in an effort to sustain and, eventually, restore structure and productivity. There were advertisers who emphasized creativity, some who emphasized technology; others who sought refuge and success in continued paths of mergers and associations; and those who distinctively turned to culture and consumers for solutions. Still others emphasized philanthropy. They asserted that advertising's woes stemmed from the lack of conscience within capitalism. There were also those who moved from gimmick to gimmick, rising and falling with the tides of change. Others returned to interdisciplinary efforts and full-service agencies hoping to rebuild a mass audience once more. Further still, many advertisers advocated independence through small agencies and boutiques that catered to niche audiences. They vehemently affirmed that large general markets were a thing of the past. Overall, there were neither clear solutions nor decisive directions.

Of particular significance to the impact of convergence on advertising were trends in the areas of branded entertainment and measurement. Branded entertainment symbolized the complexity of contemporary intermingling represented through convergence. Furthermore, branded entertainment manifested the evolution and depth of the intrinsic relationship between advertising and popular culture. The emphasis on measurement stemmed from advertising's technological environment and media climate. Severe economic losses made accountability a huge priority throughout the industry (Inukonda & Pereira). Additionally, technology increased the ability to measure media at an unprecedented level. Since the bottom line of advertising is most often the sell, understanding measurement became key to monetary efficiency and the restoration of the discipline.

The widespread usage of branded entertainment throughout the advertising industry resulted in more integrated tactics and strategies. As indicated by advertising's futile attempts to persuade and influence generation X, the youth generation of the 1990s, consumers had grown tired of advertising. Traditional advertising tactics declined in effectiveness. Increased options, media fragmentation, and market saturation further compounded matters. As a result, advertising sought effective ways to connect with audiences and create resonant relationships between brands and consumers. Branded entertainment was used as a method to produce these objectives while entertaining audiences in a less intrusive manner. Branded entertainment had to align perfectly with the brand and deliver a powerful seamless message across media platforms. It appeared in the forms of sponsorship, product placement, events, and programming.

Branded entertainment was used throughout the advertising industry to help offset the impact of convergence. It was highly successful among advertising conglomerates and was able to leverage relationships across industries and media to form powerful partnerships. In 2002 conglomerates desired to shift consumers' views of advertising content. Conglomerates wanted consumers to perceive advertising with the same regard and enthusiasm as they did for their favorite entertainment programming. Taylor affirms that insiders in every corner of the industry seemed to be in agreement that convergence was occurring. It was apparent that a crisis had engulfed multiple industries including music, film, gaming, paraphernalia, broadcast, and confectionery.

The music industry suffered devastating losses due to advances in technology. The Napster file-sharing controversy, which led to copyright infringement, legal action, and economic calamity, poised the music industry for partnership to recoup losses (Inukonda & Pereira). A number of conglomerates believed that it was best to work together and strengthen alliances in efforts to combat economic turmoil and rebuild a mass audience. In regard to its advertising, John Hayes, the chief marketing officer at American Express, stated, "We have moved out of the buying world and entered the world of content and channel integration in a significant way" (Taylor, p. 407). During a meeting of advertising and content providers, an advertising professional remarked:

> We're not in the business of content or commerce but in creativity. We are in the business of creating brand experiences. Brands are the central focus of what we do. Our industries are moving independently but in the same direction. (Taylor, p. 407)

The relationship between Michael Jackson and Pepsi set a precedent. It demonstrated the enormous power and profit that can be attained through convergence. A contemporary model for decisive endeavors of content and commerce was the appearance of rock star Sting in a commercial for Jaguar in 2000. Although Sting's popularity had waned, he possessed a loyal fan base and qualities that advertising could exploit. Sting was slated to release his debut video, "Desert Rose," for his upcoming album *Brand New Day* and was seeking to generate buzz. "Desert Rose" contained untapped resources, and Jaguar was utilized as the vehicle of choice within the video. Miles Copeland, Sting's manager, realized that this could present a major opportunity for both industries with potential success.

Copeland sent the video to Jaguar's advertising agency, Ogilvy & Mather, stating his offer of a deal: "If you will make the video look like an ad (advertisement) for my record, I'll give it to you free" (Taylor, p. 409). Copeland asserts that this was the first time that a known artist had promoted an unknown song in this capacity. Jaguar's worldwide director of sales and marketing affirmed, "Once we saw it, we realized the enormous opportunity to produce a television advertisement using footage from the video" (Taylor, p. 409). In a press release, Sting stated, "The director proposed a number of cars to be used in the video and I chose the Jaguar S-type. It's a beautiful car and it evokes the feeling of style and success we were trying to achieve" (Taylor, p. 409).

In the final agreement, Sting appeared without a fee in return for the use of excerpts of the video, "Desert Rose," in the commercial for Jaguar. In 2003 *Advertising Age* lauded this intermingling of content and commerce. The publication contended that the collaboration between Sting and Jaguar served as a model for global industry (Donaton, 2003). The song received little airplay before the commercial was broadcast. The projected sales for the album were estimated at about a million copies. However, after the commercial aired, sales soared. Sting's album, *Brand New Day*, became the bestselling solo album during that time, with 4 million copies bought in the United States alone. Moreover, Jaguar enjoyed a surge in sales (Donaton).

The successful campaign symbolized a growing trend throughout the music industry. Convergence and the intermingling of media and advertising had become so pervasive that previous standards of success had become virtually irrelevant. Radio airplay was no longer deemed as credible. Moreover, music sales, popular music charts, or signing a contract with a major reputable record label had also diminished in importance. Consequently, most musicians could no longer rely on radio airplay to promote their work. Instead, the music industry increasingly turned to the advertising industry to place their music into commercials, television programs, and films. The reproach surrounding music commodification virtually disappeared. The executive creative director of Deutsch, Los Angeles, affirmed that the biggest change within the industry was its willingness to apply unconventional approaches and strategies. Deutsch affirmed that the sell-out stigma attached to creatives was gone. Peter Nicholson, a chief partner at the advertising agency Deutsch, New York, contended:

> The old cliché' that the artist *sold out* doesn't apply in this situation, because it is a harmonious relationship that is built on the truth of popular culture's perception of the music and the brand. The music is cool. The brand is cool. And both can become part of the DNA of how a person defines him or herself. (2007, p. 4)

Success was also measured through branded entertainment. A music executive boasted to a music producer at a prominent advertising agency of his client's phenomenal year, which included a *Grey's Anatomy* television show placement, two advertising placements, and a potential film license. Josh Rabinowitz, a senior vice president at Grey Worldwide, asserted:

> More artists are going to be broken [i.e., introduced to the public] through corporations, with the agencies as talent scouts ... The agencies are kind of like the A&R [Artist and Repertoire, a function once assumed by record companies], and the client's blessing is the green light. My theory is that sooner or later, the record companies will be cut out of part of the process. (Lindsay, 2005, p. 26)

By 2007 convergence had become readily apparent. Peter Nicholson concluded his opinion piece written in *Billboard*, a premiere American trade publication, as follows:

> I will end on my bias as to why an advertising agency makes for a great partner if you are an independent band ... Most agency creatives are artists at heart. And in some agencies, they actually get to be more artist than marketer. Creatives spend a lot of time making ideas that take on a bit of their own personality. So the work becomes personal and not commercial. Or, as I like to say, a lot of care has gone into the work. The creatives share the same understanding that any artist has: your work is precious and it is personal and must always be respected. (Nicholson, 2007, p. 4)

Although, creative practitioners have long since used the trope "creativity" to describe their work, Taylor contends that this terminology has now become a vital component within intentional lexicon and rhetoric associated with advertising. Taylor argues that the term is applied today as a signifier of creative practitioners' natural affinity with the entertainment industry. It connotes power, confidence, and ability by indicating that advertising creative practitioners are poised and capable of any job within creative industries.

Undoubtedly, there is authenticity surrounding this terminology. Advertising agencies have been expanding for decades. Career paths and job responsibilities have had to adjust with the expansion of the industry. However, this marked occasion indicates that this rhetorical strategy has been used to gain traction outside the world of advertising, symbolizing the underlying presence of convergence.

After the economic calamity associated with the dot-com bust, the advertising pendulum swung back to research, testing, and measurement. The discipline recognized that a sophisticated approach was necessary. Traditional advertising media buys were now considered inefficient and resulted in financial loss and misappropriation. Pricing strategies were often the same regardless of consumer-viewing behavior (Tungate). Successful Internet pricing strategies provided seminal models for advertising and promised hope for restoration and efficiency.

Although the promise of renewal had mass appeal, many believed that placing an emphasis on measurement was far too great a price to pay. Due to measurement's intricate link with research, it did not afford the synergy or momentum many believed was necessary for overall success. This became yet another source of contention within advertising.

Creative advocates argued that advertising's strength in times of turmoil was often found in creativity and risk taking. Advertising reveled in its ability to connect and evoke emotive behavior from consumer audiences. Some advertising professionals warned that an overreliance on measurement would strip advertising of its essence. Moreover, they argued that those who reduced advertising to measurement failed to grasp the complexities of the institution itself. Although the bottom line was to sell, advertising as an institution in a capitalist market served a much larger purpose that often could not be quantified. Raymond Williams, cultural critic and theorist, expanded upon this topic in his seminal article, "The Magical System":

> It is impossible to look at modern advertising without realizing that the material object being sold is never enough: this indeed is the crucial cultural quality of its modern forms. If we were sensibly materialist, in that part of our living in which we use things, we should find most advertising to be of an insane irrelevance. Beer would be enough for us, without the additional promise that in drinking it we show ourselves to be manly, young in heart, neighborly. A washing machine would be a useful

machine to wash clothes, rather than an indication that we are forward looking or an object of envy to our neighbors. But if these associations sell beer and washing machines, as some of the evidence suggests, it is clear that we have a cultural pattern in which the objects are not enough but must be validated, if only in fantasy, by associations with social and personal meanings which in a different cultural pattern might be more directly available. The short description of the pattern we have is *magic*: a highly organized and professional system of magical inducements and satisfaction, functionally very similar to magical systems in simpler societies, but rather strangely coexistent with a highly developed scientific technology. (1980, p.176)

Advocates of measurement retort that history has also revealed that all markets and exchange evolve toward efficiency, transparency, and accountability. They contend that such a direction will eventually benefit both the principal buyers and sellers. Measurement supporters argue that this is not merely a matter of opinion but rather a fact. When properly enabled by technology, both the online advertising market and the broader advertising market will move in this direction. Furthermore, proponents of measurement proclaim that companies that are compelled to spend on advertising to maintain competitive advantage in perplexing markets will not continue this practice without assurance. Companies will eventually require confidence in the return on their investment through a specific and quantifiable source in both online and traditional media outlets.

Magic and efficiency seemingly converged with the revolutionary election of America's first African American president, Barack Obama. Moreover, the Obama campaign provides a powerful demonstration for the uncertain role of advertising agencies within the future of the profession. In a nation highly characterized by hypocritical actions and endemic racism, Obama's election was a groundbreaking event. Merely 43 years after African Americans had achieved legislation for civil liberties, including the right to vote, an African American male was elected to the nation's highest office in 2008. Although his political platform featured major issues, notably the Iraq war, universal health care, and the great recession, Barack Obama's 2008 presidential campaign was distinguished by its exemplary use of advertising tactics and its ability to successfully harness new media.

Unlike presidential elections of the past, advertising agencies were largely absent from this campaign. In a growing trend, media companies began working directly with clients to achieve efficiency and speed. The new media environment increasingly required that work be produced in real time. Although this rapid pace created opportunities for concurrent testing, it inhibited prolonged approval processes. As a result, advertising strategies that involved free "earned media" gained momentum and made media experts highly desirable among clients. Communication methods that reached audiences and spread content through Facebook, YouTube, Twitter, press coverage, and mobile technology created added value for clients and stretched shrinking budgets. However, for many, advertising agencies and advertising practitioners' strategies such as these were synonymous with more work and less money.

As demonstrated by the 2008 presidential election, advertising techniques were highly evident but the force that drove the campaign was media. This media prowess was achieved by the digital advertising agency Blue State Digital. Integrated communications were used to communicate across all media platforms. Branding united the campaign through the use of iconic visual imagery and resonant slogans. The campaign rhetoric harnessed the use of popular culture tropes and universal truths: family, love, struggle, identity, and triumph. The campaign promoted the ideology that American people are decent and generous, with desires for civic engagement. Like advertising, Obama's 2008 presidential campaign largely appealed to youth audiences and created a sweet spot among baby boomers. Millennials were empowered and boomers were given the opportunity to expand upon many of the social justice issues that characterized their youth.

In addition to a pertinent target audience, success was achieved through Blue State Digital's ability to leverage culture through an acute understanding of how media characterize society. The media environment during these times was characterized by empowerment, real time, communal experience, and participatory culture. It was representative of bottom-up, rather than top-down, communication flows and processing. Consequently, Blue State Digital was able to resonate with the public through the use of highly relevant rhetorical strategies. Blue State Digital also provided experiences to harness communal tendencies and participatory culture. The campaign made use of grassroots tactics. Extensive fundraising efforts were accomplished through small donations, volunteerism, and opportunities for personal experiences with the candidate. Obama's 2008 presidential campaign allowed constituents to be a part of a larger communal experience through social media. This connection is deeply rooted in an oral tradition that elicits experiential connectivity through ritual communication methods (Carey, 1989).

Media understanding was further demonstrated through groundswell efforts that used social media networks, including Facebook, Myspace, and interactive Web 2.0 platforms. Chris Hughes, a member of the millennial generation and cofounder of Facebook, was one of the campaign's key strategists. Hughes developed the Obama campaign's highly effective web blitzkrieg, an aggressive Internet strategy that facilitated interactivity through linking social networking, podcasts, and mobile devices. Like many groundbreaking creative ventures of the past, a larger advertising conglomerate, WPP, purchased Blue State Digital in 2010 (Hall, 2010).

The success of the 2008 election informed many of the decisions for Barack Obama's presidential reelection campaign in 2012. Again, the campaign primarily worked directly with media experts and circumvented advertising agencies. Facebook was incorporated in sophisticated and innovative ways through the use of targeted sharing. Strategists employed methods that demonstrated the power of influence. In contemporary culture, messaging and content from "friends" is often more effective and influential than messaging from advertising campaigns. Thus, effective "digital volunteerism" facilitated the assistance of millions of casual and committed supporters and increased potential constituencies. Strategists employed methods that lowered barriers for contacting voters, while simultaneously fortifying the influence of messaging. Moreover, hundreds of thousands of people utilized the campaign's Facebook applications to influence their networks. Campaign manager Jim Messina affirmed:

> What Targeted Sharing was—and I think it's one of the most important things we did—was a Facebook app that allowed you to go and match your Facebook world with our lists, and we say to you, "Mike, here are five friends of yours that we think are undecided in this race. Click here to send them a piece of viral content. Click here to send them a factsheet. Click here to ask them to support the President." ... [I]t took us a year of some amazing work of our talented technology team to figure out how to do it. But we were able to contact over 5 million people directly through their Facebook worlds, and people that they knew. (Allen, 2012)

The elimination of advertising agencies in the advertising campaign process has been reinforced through measureable success. This is in stark contrast to successful presidential campaigns of the recent past. For example, in 1992, the presidential election campaign of Bill Clinton achieved major success through the efforts of prominent Madison Avenue advertising agencies. Donny Deutsch, creative director of Deutsch, propelled Clinton forward through his bold use of media and creative-branding techniques. Deutsch, an influential veteran of the advertising industry, had achieved compelling success with notable brands, including IKEA, Johnson & Johnson, General Motors, DirecTV,

Expedia, Mitsubishi, Revlon, Bank of America, and others. Deutsch employed strategies that divided the creative team into separate units to produce positive advertisements, attack advertisements, and response advertisements (Kolbert, 1992). Television was an essential media involved in the campaign strategy. This approach was influenced by the model of success achieved by Ronald Reagan in his presidential election campaign of 1984. Advertising creative giant Philip Dusenberry of BBDO successfully spearheaded Reagan's presidential election campaign, which emphasized television advertisement. The campaign was made famous by the "Morning in America" commercials. BBDO has a history of major influence in advertising campaigns, including elections ranging from Calvin Coolidge (1924) to Richard Nixon (1956).

Presidential campaign elections provide powerful exemplars for the characterization of crisis demonstrated through client relations in advertising. Contemporary advertising reveals a significant decline in the relationships between advertising agencies and advertisers. This places the role and functions of advertising agencies in peril. The title and role of "agency of record" have become nearly obsolete. Many large brands and influential clients are opting to work with specialized media boutiques as they are increasingly fearful in an era characterized by convergence.

Technology has intensified many of these concerns. Digital firms have created applications that digitize and automate inherent functions of advertising agencies, which have further pushed advertising agencies toward the margin. Examples include MediaMath and DataXu, which have contributed platforms to automate media and advertising buys; buildabrand.com, which has reduced the craft of branding to customized algorithms; and Lotame, which has created applications for databases to ensure audience data management (Sacks, 2010).

Although technology has exacerbated concerns among advertising agencies, there are those that advocate its use for measurement. Successful pioneers of the dot-com era have provided effective models for technology-driven measurement practices. eBay and Amazon have served as inspiration for those seeking to harness the power of the Internet for promotional, advertising, and commerce endeavors. Amazon leaders in particular adopted technological innovations that replaced mundane tasks once routinely handled by labor-intensive staff. Business operations were efficiently streamlined. Amazon's Internet commerce strategy became known as clicks and mortar. It entailed a commonsense melding of solutions that were fast, seamless, and slick. Moreover, Amazon's ecommerce solutions corresponded with traditional services and distribution to ensure viable consumer experiences (Simon & Joel).

As a pioneer, Amazon's legacy dictates that companies pay close attention to the appropriate and fluid nature of the relationship between content and medium. Jeff Bezos, American entrepreneur, founder and CEO of Amazon, first concluded that the Internet was the only place that could house the world's largest bookstore. Aside from establishing the appropriateness of this union, the key focus was creating a mass audience and ensuring an ideal consumer experience. Amazon harnessed advances in technology, such as the silicon chip and encryption, to address the Internet's credibility issues and ensure the privacy concerns of its patrons. In its early stages, Amazon intentionally generated no profit. In fact, the company's Get Big Fast strategy kept pricing low and ensured a huge mass market, which was prioritized over profit (Brandt, 2011). Bezos applied his digital savvy to traditional media with his multimillion dollar purchase of the *Washington Post*, officially ending 80 years of local control of the newspaper by the Graham family (Pew, 2014).

The colossal success of Amazon spawned legions of imitators, as the advertising industry quickly adopted similar strategies in hopes of achieving the same levels of success. Another company that generated comparable parallels of emulation was Google. Much of Google's accomplishments were embedded within utility, relevancy, and constant innovation. Its prominent objective lay within the ability to organize an immense amount of information made accessible by the Internet. Simultaneously, it made the information universal and accessible to the public. Organization was achieved through search engines that estimated the importance of websites based on incoming links. Google's primary multibillion dollar revenue was generated through advertising, which was based on successfully redefining the practice of targeting within the information era (Simon & Joel).

Google quickly established the industry standard in contextual targeting, which is based on the belief that the context of a website is most important in determining what advertising messaging should be displayed. Thus, when a user navigates online, advertising will appear that correlates with the content of each particular site. Google efficiently accomplishes this task through its programs, AdSense and AdWords. Advertisements are administered, sorted, and maintained by Google. Revenue is generated either on a per-click or per-impression basis. Marketers are offered free website visitor statistics through Google's analytics service (Simon & Joel).

Google attempted to keep pace with constant advances in technology by organizing information in accordance with images, location, language translation, and videos. It improved on Internet productivity through services provided by Gmail, Google Docs, Google News, and cloud services. Google also has introduced an open source web browser, Google Chrome, and the mobile operating system, Android. It also has ventured into social networking through Google+. Ken Auletta, author of *Googled: The End of the World as We Know It (2009)*, affirms that Google has achieved in 6 years what it took Microsoft 12 years to accomplish. Although Google has achieved a great deal and expanded in a number of areas, it proclaims to be, at its core, an advertising business.

Critics doubt that this proclamation has any validity. In fact, over the years Google has encountered harsh criticism of its business models and emphasis on measurement. One such argument asserts that Google is simply an advertising industry disruption built upon a technology-driven market reorganization:

> Google engineers ... have no way to quantify relationships or judgment. They value efficiency more than experience. They require facts, beta testing, mathematical logic. Google fervently believes it is shaping a new and better media world by making the process of buying advertising more rational and transparent. In its view, the company serves consumers by offering advertising as information. (Inukonda & Pereira, p. 7)

Google's critics contend that advertising has never been expressly about efficiency and inventory. Advertising has never allowed technology to solely define its ethos as an industry. Technology is often best described as a system to deliver media and culture to users. In this regard, the state of crisis within advertising could be more easily understood from a cultural standpoint rather than primarily the technology in which it is characterized.

Like the unrest surrounding branded entertainment, the contention encompassing advertising measurement may not be settled within the near future. In addition to contrasting discourse, these issues face contingencies often based on market accelerants. Broadband advances and accessibility, as well as market competition, create potential roadblocks to resolution. Convergence has ensured that stakeholders will be dynamic, complex, and constantly in flux. New contenders, viable technologies, and pricing models are emerging continually. At the inception of online advertising, one of the primary pricing models was CPM (cost per thousand). However, following the success of Amazon and Google, numerous pricing models have emerged that offer more efficiency for advertisers and allow for a larger margin for publishers, including DoubleClick, ad networks, video ads, in-game advertising, and mobile networks (Simon & Joel).

One of the most significant advances to take place post the dot-com bust has been the genesis and explosive growth of social media (Hendler & Golbeck). Social media emerged in the world of Internet advertising in 2004 and quickly soared in popularity. With the advent of social-publishing platforms and social networks, namely Facebook, YouTube, and Twitter, this new technology began to generate significant traction. In less than a decade, Facebook has amassed 500 million members; more than 14 billion videos are viewed each day on YouTube; and Twitter has more than 165 million users (Yeomans, 2010). Again, the potential promise of a mass audience has lured many advertisers into the world of social media.

According to recent Nielsen data, 64% of marketers planned to increase their social media advertising spending by 2013 (Stambor, 2013). However, there is little consensus on how marketers will determine the value of their return on investment for social media advertising. A great number of advertisers affirm that the primary purpose of social media advertising lies within branding or raising awareness. Accordingly, metrics, including Facebook likes, Pinterest pins, and click-throughs, have been utilized to gauge the success of advertising campaigns. However, clients have indicated that the metrics they most prefer are those that would generate sales and create brand lift (Stambor), yet these systems for measurement have yet to be established.

Contemporary advertising is plagued with a number of issues that inhibit the development of sound metric systems. Technology has increasingly become an evasive target. Rapid progression has led to shortened lifespans for various media technology. To make matters worse, modern consumers are migratory and demonstrate very little loyalty. In some instances, technology has become antiquated and consumers have migrated elsewhere prior to the development of measurement systems. Nonetheless, the potential to connect and influence mass audiences through social media has offered significant promise.

Facebook introduced in-feed advertising. These advertisements appear directly on Facebook users' newsfeeds and allow brands to directly target consumers in a customized, less intrusive manner. Advertisers encountered click-through rates on Facebook that delivered favorable returns on investment. Additionally, Twitter launched its own self-serve advertising platform in 2013. It provided the opportunity for celebrities, thousands of small businesses, and personal accounts to promote tweets. Moreover, Pinterest, a popular visually driven content-sharing social network, became a massive hit with retail brands and users. Although Pinterest has yet to develop an effective method for monetization, it has adjusted its aesthetic and organization to benefit brand influence and generate more retail interest (Stambor). Furthermore, BuzzFeed, a popular digital news organization, has made native advertising a pillar of its financial strategy. It leverages its native advertisements through purchasing ad space on social-networking sites to drive users to advertiser content on BuzzFeed.com. In 2013 BuzzFeed created an advertising network, pitching stories to advertising agencies to run sponsored posts on other websites' homepages (Pew).

Technology has pushed contemporary advertising in varied directions. However, strict adherence to the demands of a rapidly progressive technological environment may shroud other major controversies within advertising. Unresolved issues remain ever present.

Factions within advertising are a major source of controversy. Factions contribute to ineffective communication, which has led to the exacerbation of other major issues. Advertising silos prevent problem solving, understanding, and collaboration within the discipline. All the while, empowered consumers have increasingly gained options and access to more information than ever before. Consumer audiences have undergone a revolution, while advertising has not.

Technology also has deeply influenced advertising agencies, consumers, and advertisers. Corporations have become increasingly transparent. Corporations have become increasingly defined on their actions and interactions with consumers rather than through controlled imagery or personas created through advertising. Additionally, trends involving grassroots movements have prompted the use of consumerism for philanthropic or service-oriented endeavors. Advertising campaigns, produced by independent agencies, boutiques, entrepreneurs, and consumers alike, have gained traction by facilitating the use of service, utility, or philanthropy to engage audiences.

Influence is not solely coming directly from agencies and advertisers and then being dispersed throughout the public. Recent trends indicate that user-generated content and grassroots movements have helped shape the advertising agenda. Through a bottom-up process, several models have emerged that generate the potential for emulation among conglomerates. Prior to the information age, linear communication precipitated ideation from the top down. However, contemporary technology has empowered a more communicative process that involves an interactive and circular conversation as opportunity to create solutions for a multitude.

Potential solutions that have emerged in advertising indicate that the new paradigm will not be based solely on market efficiency. In order to restore the productivity of the entire discipline, there must be value exchange and consensus among all stakeholders. Audiences now have a powerful voice and ways to influence brands and other consumers. In order for consumers to continue to engage with brands, there must be a reciprocal exchange in which both parties achieve their wants and needs. Models suggest that advertising networks and campaigns will supply market-driven platforms for this emergent system of value exchange. Technology suggests that multiple voices will be heard within this circular conversation. Communities and culture have come to embody the institution of advertising. Undoubtedly, greater accountability will be required of advertising. However, potential realignment may foster the magic that will restore advertising to greatness well beyond its golden age.

PART II
TRIADIC CONVERGENCE, INSIGHTS, AND IMPLICATIONS

· 6 ·

OVERLAPPING PHENOMENA

Triadic Convergence

The concept of convergence was introduced to mass communications discourse in the 1980s through the works of Ithiel de Sola Pool. In *Technologies of Freedom* (1983), de Sola Pool explores how advances in technology impact society, policy, and freedom. He identifies a convergence of modes that can eradicate the boundaries of communication established by the market and government legislation.

In 1980 former chairman of the Columbia Broadcasting System (CBS) William Paley affirmed the prescience of convergence within industry. Paley stated that while corporate ventures were preoccupied with establishing boundaries and defending territory, the extent to which they were being drawn together by the vast revolution in electronification—innovations that allow pulses of electromagnetic energy to embody and convey messages—was ignored (de Sola Pool).

Nicholas Negroponte further explores insights associated with convergence. Negroponte was among the first scholars to introduce convergence into the lexicon of popular culture through his writings in *Wired Magazine* and *Being Digital* during the 1990s. The intrinsic relationship between popular culture and advertising facilitated a greater acceptance of this phenomenon. However, conceptualizations of convergence were often broad and indirect, which contributed to obscure interpretations.

In contemporary discourse convergence is most commonly used to identify the phenomenon of the integration and merging of media, technology, and culture (Danesi, 2012). While difficult to map or to locate, and fluid almost to the abstraction, convergence is a complex mechanism that spans technologies, economics, social organizations, cultural frameworks, global networks, and property configurations (Guertin, 2012).

Although extremely valuable, such characterizations of convergence are not specific to advertising. A conceptualization of convergence more closely related to advertising promotes deeper understanding and contributes to restoration of the discipline or progression toward a revolutionary paradigm shift. Hence, an isolation of the phenomenon within the historical context of the intrinsic relationship between advertising and popular culture offers tremendous insight.

Moreover, convergence has rapidly evolved within recent years. Increased intermingling has led to complex outcomes characterized by the unique interactions between convergence and specific institutions. Thus, for discussions involving the institution of advertising, convergence is conceptualized as "triadic convergence," a complex and dynamic force comprised of the sophisticated intermingling of three core elements—media, technology, and culture—is offered. Intermingling is characterized by the constant mutative and adaptive synergy achieved among media, technology, and culture. This interwoven relationship is shaped and shifts in accordance with the characteristics of the shifting locus of power within the triad (see figure 6.1).

The model shown in figure 6.1 is a departure from previous interpretations of convergence. Prior models depicted four circles moving toward each other (Brand, 1987; Wirtz, 2001). The circles typically represented information technology, telecommunications, media, and consumer electronics. As suggested by previous models, industries were predicted to merge. Thus, boundaries were expected to erode and American industry would function as a seamless entity.

Years have proven this conceptualization to be inaccurate and oversimplified. The model for triadic convergence specifically addresses complexity. Triadic convergence depicts three moving circles united to represent a singular, yet dynamic, force. Each circle is dimensional and represents a distinct locus of power within the triad. No circle rests on an axis. Rather, each circle

Figure 6.1. Triadic Convergence Model.

overlaps and intersects in a symbolic representation of the complex intermingling of media, technology, and culture.

In this depiction of triadic convergence, media is positioned at the top. This positioning represents centralized power that is characterized by widespread conglomeration. Technology and culture are positioned opposite each other at the base of the model. This positioning is symbolic of the continuous relationship between technology and culture in which the cause-and-effect are sometimes reciprocal and not fixated. The model also includes dimensional arrows, which represent the unpredictable nature and mutative intermingling inherent to triadic convergence. It is important to note that the positioning of the three elements—media, technology and culture—is not permanent and will shift as change occurs.

In regard to advertising, it is suggested that triadic convergence is not overly identified within the context of its comprising entities. Impressions of triadic convergence are generally associated with utopian claims or dystopic admonitions based on extrapolation from technology. A deeper analysis and understanding are needed (DiMaggio, Hargittai, Newman, & Robinson, 2001). Triadic convergence can be understood as a fluid and intersecting set of forces that is accompanied by practices, technologies, events, and complexity.

Electronification is the force that propels triadic convergence. Electronification has come to embody the universal signifier of the premier driving force behind the new-wired world and the revolutionary change that accompanies it. Electronification is the control of the flow of electrons that facilitates pulses of electromagnetic energy to embody and convey messages. Its importance cannot be overstated, especially because electrons were not always subject to a great degree of control. Technological progress has facilitated the ability to manipulate, store, amplify, and transform electrical signals. Transmission from sender to receiver now has the ability to flow in digital codes (Negroponte, 1995). Consequently, in every medium the manipulation of symbols in computers and the transmission of those symbols electrically have the capacity to be utilized at crucial stages in the process of production and distribution. Moreover, electronic methods have proven far superior to their predecessors (de Sola Pool).

The preeminence of electronification has drawn attention to the works of media and communications scholar Marshall McLuhan. In the 1960s McLuhan wrote that the dominance of visual culture would eventually evolve toward a more eclectic interactive experience. In what McLuhan termed "electronic interdependence," he contended that electronic media would replace visual culture with oral culture. McLuhan claimed that in this new age spurred by electronification, society would shift from individualism and fragmentation toward a collective identity (McLuhan, 1962). McLuhan's coined this new social organization the "global village."

Increasingly, McLuhan's (1962) predictions of a global village have become seemingly apparent. In fact, they foreshadowed the triadic convergence revolution through electronification, which facilitated the creation of a vast collection of computer networks fused to form a single entity—the Internet. Over time, the Internet has come to embody a compact network for global transmission of data across time and space (Alkalimat, 2004). Much like Dewey's (1927) ideals of the communicative process, the Internet has expanded conversations, created shared experiences, and contributed to unique behaviors for billions of users worldwide.

In addition, the Internet has also helped to realize McLuhan's assertions that electronic communications media are extensions of the human mind and senses that cultivate communal interactivity. This realization is apparent with communication technology that is not simply a structural system for usage and gratification but also a force that yields enormous cultural ramifications. Communications scholar Harold Innis posits that communications technology has a significant influence on how society and culture are constructed and affects the development of subsequent media and technology. The concentration on a medium of communication implies a bias in cultural development and social organization. Accordingly, each major period in history takes its character from the medium used most widely during that time (Danesi). Consequently, contemporary culture is characterized as highly participatory. This is in sharp contrast with previous communication models, which suggested paradigm shifts.

It is within the current communication model—the Internet and concomitant components—that the triadic convergence revolution finds currency and relevance.

Advertising

In addition to triadic convergence, advertising is the other phenomenon involved in the impending paradigm shift. The significance of a phenomenon, as with any institution, is largely based on the relationship achieved among the interaction of several dimensions both within and external to the industry (see figure 6.3). Therefore, the interdimensional dynamics, structure, and nature of advertising are essential to understanding the impact of triadic convergence.

Given its dense complexity, incessant metamorphosis, as well as its economic and cultural significance, the meaning of advertising is neither simple nor succinct. It is not detached, but rather deeply connected to the world it creates. Advertising is most commonly defined as paid, nonpersonal communication delivered through various mediums with the intention to inform or persuade members of a particular audience. Advertising messaging from business firms, nonprofit organizations, or individuals are typically identified in the promotional messages delivered to audiences (Krugman, Leonard, Watson, & Arnold, 1994). The advertising industry is the systematic aggregate of manufacturing or technically productive enterprises that utilize persuasive communications to promote the sale of specific commodities, ideas, or services (Danesi).

The conceptualization of advertising as an institution is attributed to the works of mass communications scholar James Carey (1960). Vincent Norris (1980) furthered these concepts with the addition of criterion. Norris asserts, as an institution, advertising embodies the following essential characteristics:

Figure 6.2. Institutional Advertising Model.

1. Advertising orders human relationships into roles.
2. Advertising regulates the distribution of society's essential resources.
3. Advertising is ubiquitous.

The ubiquitous nature of advertising is virtually beyond question. It is estimated that the average American is exposed to nearly 3,600 advertising messages daily (Jhally, 1997). Advertising continues to proliferate as media, technology, and businesses evolve. Corporations alone invest more than $600 billion annually in advertising budgets within the global economy (Danesi).

There are numerous ways in which advertising orders human relationships into roles. Advertising conveys messaging, responsibilities, and imagery that uphold standards of influence, value formation, concepts of normalcy, and structural roles. This is largely accomplished through advertising's inextricable relationship with popular culture. This relationship creates a powerful channel to cultivate ideals while remaining deeply entrenched within society and daily life.

As an enterprise advertising orders the roles of seller, retailer, wholesaler, and target audience. Within agencies the advertising industry orders professional roles such as account executives, copywriters, graphic designers, research planners, and media buyers. Advertising also teaches people how to participate in consumer culture. Through advertising, consumers learn to understand brands, positioning, purchasing behavior, and lifestyle.

Advertising regulates the distribution of America's most valuable resources in several ways. Modern advertising has evolved from the need to sell abundance under the capitalist economic structure of the United States (Potter, 1960). Advertising stimulates the demand for goods and services so that consumers will purchase product; hence, money flows from consumers to producers. Furthermore, advertising establishes the means for manufacturers to set prices for products. The advent of national advertisers created a demand for a brand that then allowed its manufacturers to set price regulations for retailers (Strasser, 2004). Through advertising's relationship with media, advertising greatly influences the content that is delivered to audiences. Consequently, media are heavily reliant upon advertising for revenue. Advertising's relationship with media reinforces fixed structures for the regulation and distribution of resources.

Due to its many roles and functions, advertising is a complex and evolving phenomenon. The character of advertising is often dependent upon the character of market structure and the values and beliefs that support that structure. Accordingly, developments in culture, changes in technology, or changes in the location of economic power (as currently depicted in contemporary media) will greatly affect what advertising is and does.

As an institution, advertising is an "instrument of social control," affecting society by guiding the lives of the masses generally and individuals specifically. This is achieved through conceptualizations of distinct behavior. Accordingly, individuals and society are encouraged to conform to conceptualized behavior in order to establish norms of conduct. Such norms of conduct protect against chaos and ensure stability and the organization of economic activity. Thus, advertising will primarily view members of society as consumers. This assertion creates a potential area of conflict within contemporary advertising and periods characterized by triadic convergence.

Typically, power rests with producers, not consumers. Advertising typically assumes the role of producer. Yet, the information age has dispersed power among consumers through decentralized technology. Moreover, triadic convergence has contributed to instability. In addition to its elusive and intermingled nature, triadic convergence has caused the disintegration of organization surrounding economic activity through massive redistribution of media. Thus, the question arises: With the distinct understanding that knowledge is power, what is the relationship between information and knowledge?

If citizens perceive information primarily as entertainment, then citizens can be perceived as consumers. If citizens perceive information primarily as knowledge, then citizens can be considered producers and, therefore, empowered. Consequently, trends in advertising that emphasize entertainment and humor may be influenced by consumer demands or by attempts to maintain the status quo. Such instabilities within power relationships represent upsets within norms of conduct, thus signaling an impending paradigm shift.

Due to triadic convergence, significant change has altered advertising causing many to contemplate its future. Over the years advertising has become a revered institution within the economic operation of global order and the core of societal functions. The indispensable role of advertising to the economy and culture contributes significant doubt as to whether or not this industry will cease to exist.

Overlapping Phenomena

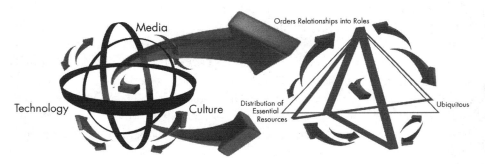

Figure 6.3. Impact of Triadic Convergence on Advertising.

Both triadic convergence and advertising are evolving multifaceted phenomena. A discussion of their significance requires consideration of concomitant factors (Kenway, 1996), none of which exist in isolation but rather in contiguous, interwoven, overlapping, and seemingly contradictory capacities.

The impact of triadic convergence on advertising has contributed to several major occurrences. Among the most prominent is the impending paradigm shift. Therefore, it is useful to understand the parameters of paradigm shifts as they specifically relate to contemporary advertising. A paradigm shift is a change in the governing principles of a discipline. It occurs when a discipline collectively adopts a new pattern or structure in order to restore productivity, which is restored through solving the most pressing issues of the discipline. Paradigm shifts often relate to a phenomenon that elicits an unforeseen modification in worldviews, resulting in a blurring of practices and ideologies (Kuhn, 1964).

Paradigm itself is a dimensional concept that often requires explication. "Paradigm" is defined as the entire constellation of beliefs, values, techniques, and so forth shared by the members of a given community. Paradigm can also refer to one sort of element within that constellation that can replace explicit rules as a basis for productivity (Kuhn). A paradigm is a view of the world that successfully informs the theories and practice of a profession. Paradigms are often associated with major occurrences that inspire a new way to see the world.

Paradigms are of vital importance to the progress of a field. They structure the set of beliefs, assumptions, and values that unite a community and facilitate the undertakings of their enterprise (Drønen, 2006). Paradigms create meanings for terminology that characterize the discipline. In addition, paradigms largely influence how practitioners and scholars view subjects within their craft. Paradigms determine the questions and narratives regarded as valid within a field and constitute the rules for theory, models, frameworks, methods, and research (Hoyningen-Huene, 1993).

If an established paradigm is incongruent with reality, chaos will erupt. Thus, the prescience of a state of crisis suggests that a paradigm shift is underway. The core of crisis is based on discrepancies between theory and fact. It is characterized by an accumulation of anomalies within a discipline. Anomalies are a natural occurrence. However, the accumulation of anomalies is not. The accumulation of anomalies indicates that, when applied, the accepted paradigm has continually failed to produce successful outcomes. At the stage in which there is massive accumulation of anomalies and a cessation of productivity throughout the discipline, crisis becomes universally recognized as the primary agenda of the discipline in order to restore productivity (Kuhn). Until crisis is resolved widespread confusion and chaos will persist. Turbulence and upheaval are boundless.

The modern advertising environment represents an opportunity to restructure the direction of the discipline. Although convergence is the way of the future, it is taking shape now—in essence, creating new paradigms (Jenkins, 2006). Thus, greater understanding regarding the impact of triadic convergence on advertising is needed. As issues concerning media, technology, or culture arise in advertising, triadic convergence will be further propelled into its mainstream agenda. Much of advertising's success is dependent upon its ability to harness technology. Yet, triadic convergence contributes to tension and instability in environments characterized by technological dependency. These concerns are intensified because recent advances in technology promote communal tendencies and decentralized power, while advertising conglomerations represent centralized power. Advertising has expanded its global roster; therefore, understanding is needed to create advertising solutions that resonate with diverse cultures and lifestyles. Moreover, media has a significant influence on culture and both have undergone formidable shifts in recent years.

Further understanding of this phenomenon is needed for financial success and the restoration of productivity. Convergence has greatly affected how advertising relates to audiences, measures media, and achieves commercial success (Deuze, 2007). Additionally, exacerbating conditions surrounding economic productivity are a source of grave concern. Advertising is closely linked to a number of institutions and industries that have experienced severe financial misfortune in recent years (Lee, Taylor, & Chung, 2011). Moreover, the ongoing global recession has contributed to cautious spending patterns among consumers and a considerable loss of income within affluent markets (Carmichael, 2010).

The impact of these overlapping phenomena is a major concern. It is virtually inescapable as it is comprised of elements that are at the core of lifestyle and the global economy. Awareness and deeper understanding contributes to discovering methods to restore productivity throughout advertising. Knowledge of the triadic convergence phenomenon presents an opportunity to create a more inclusive public sphere. Furthermore, these narratives help to reinforce the overall goal of this book: to increase understanding and awareness regarding convergence and opportunities presented through change.

Insights

Although a state of crisis is often represented by a number of seemingly uncontrollable factors, including downsizing, restructuring, confusion, and chaos, it exists within an ordered pattern of development (Kuhn). Furthermore, a state of crisis characterized by convergence has a distinct unfolding. Insights regarding these factors are offered to assist in trying times.

It is important to note that a number of competing factions and stakeholders have a vested interest in the resolution of crisis within advertising, thus, a host of solutions will be proposed. In the absence of a working paradigm, nearly every proposed solution will appear to be true. Since the duration of crisis is unknown, caution is encouraged. Proposed solutions must undergo effective scrutiny, sustain vigorous problem solving, and ensure collective prosperity for the future. Premature explanations could, in fact, be tactical, narrow, invalid, or contradictory (Jenkins).

Predictions contributed to the wide acceptance that formerly singular media vehicles, such as the television or telephone, would converge into a common technological base through digitization (Danesi). Others predicted the collapse of television, print, and other primarily traditional media in favor of digital technologies (Negroponte). Subsequently, many advertising agencies clamored to prematurely move clients primarily into a digital space (Tungate, 2007).

Accordingly, advertising employed similar communications models that emphasized fusion, technology, brand extensions, and industry collaborations (Tungate). As corporate giants comingled, advertising proliferated within numerous markets, including video gaming, cinema, and music (E. Williams, 2010). Collaborative efforts sought to accomplish in unison what could not be achieved in isolation (Jenkins). As a result of widespread intermingling, convergence has become much more dynamic than previously assumed. Furthermore, new media and users interact in complex ways indicating that triadic convergence is not simply driven solely by technology (Deuze). Triadic convergence has contributed to distinct behaviors and culture.

Hampel, Heinrich, and Campbell (2012) affirm that some forms of traditional advertising were prematurely discredited. They urge the industry to reexamine premium print advertising, arguing that it conveys a sense of uniqueness and prestige, which boosts consumer attitudes toward advertising as well as brands. Moreover, Nielsen research indicates that television advertising will remain a primary method for marketers to connect with audiences due to its unmatched reach and trusted value (Grimes, 2011).

Accordingly, integrated marketing communications (IMC) has emerged as a popular communications model. IMC seeks to leverage media variety by creating synergy with consistent brand messaging throughout appropriate platforms. This model has been considered a success by leading advertising professionals for its consistency, efficiency, and financial accountability (Moriarty, Mitchell, & Wells, 2012). However, its critics contend that it could be a temporary solution for a dying media that may not be sustainable in an evolving world (Bernardin & Kemp-Robertson, 2008).

Conversely, content-specific, digital interruption models have attained significant supporters due to their measurable success. Nonetheless, proponents of IMC suggest that digital interruption models are primarily successful in niche markets but lack the modularity needed for larger and more inclusive brands (E. Williams). Furthermore, others contend that the advertising industry should focus on post–purchasing behavior models and create content with which users desire to engage and interact. Avid industry leaders affirm that the current advertising landscape is not defined by IMC strategies but rather compelling multisensory ideas that ignite conversation (Bernardin & Kemp-Roberston).

Notwithstanding successful targeting of marketing and multisensual advertising models, critics declare that effective advertising decision making is situated in interdisciplinary collaboration (LaPointe, 2011). Accordingly, Wood (2012) proposes that advertising solutions may be found in behavioral, economic, and psychological models. Research has indicated that emotional engagement models trump current advertising techniques and can lead to greater effectiveness, efficiency, and measurement. Dahlen and Edenius (2007) reiterate the merit of a psychological contribution, emphasizing that cognitive processes provide valuable insights regarding consumers' perceptions as it relates to advertising credibility. Moreover, Taylor, Loiacono, and Watson (2008) suggest that convergence solutions may lie within the increased role of visual communications and graphics solutions within advertising.

Some scholars insist that market saturation and media fragmentation will not permit singular solutions in what many consider to be a waning mass market (Rust & Oliver, 1994). These advocates suggest that solutions be inclusive, fluid, and expansive to accommodate growth. Bernardin and Kemp-Robertson argue that convergence has repeatedly caused industries to shift in quick and dramatic fashions. Erevelles, Roundtree, Zinkhan, and Fukawa (2008) insist that the most effective communication models will emerge from trends and continual creative transformation. Additionally, Christodoulides, Jevons, and Bonhomme (2012) affirm user-generated content as a rapidly growing vehicle for brand conversations and advertising consumer insights. Erevelles and colleagues contend that harnessing user-generated content and consumer imagination will assist advertisers in developing effective communication models, creative strategies, and accelerate idea generation.

As characterized by a state of crisis, triadic convergence has been accompanied by a myriad of solutions that include the use of IMC, collaborative business ventures, and the restoration of traditional media. This further intensifies the need for understanding to increase clarity and decrease confusion. Moreover, advances in technology have compounded issues associated with monetization and measurement. Technology has forced old paradigms to break down more rapidly than new models that have emerged. Panic has ensued among industries, which have invested in the status quo, while curiosity has spread among those who see opportunity within change (Jenkins).

· 7 ·

TRIADIC CONVERGENCE & THE NEW MEDIA ECOSYSTEM

As a result of triadic convergence a new media ecosystem has emerged. While some elements have remained the same, many have drastically shifted. Nonetheless, it is important to understand that elements within this system are still unfolding. The new media ecosystem is a community comprised of living and nonliving components including advertisers, consumers, industries, media, and technology. This community, which is largely fueled by triadic convergence, is affected by internal and external factors.

Institutions are one of the leading external factors that affects the new media ecosystem. By and large, institutions act as a constraint upon change (de Sola Pool, 1983). Triadic convergence is a signifier of change. Triadic convergence, which is characterized by blurring, causes a redistribution of the boundaries surrounding media, technology, and culture. Thus, when an institution collides with triadic convergence instabilities erupt. Furthermore, contemporary institutions are reliant upon technology, which is also largely in flux and may contribute to worsened conditions.

Heightened levels of instability have been created by other factors, including shifts in mass communications. Like advertising, mass communications has experienced change surrounding the communication patterns and behaviors of consumers. Previously, American communication was most commonly defined by its ability to send messages for the purposes of control. Under this model, messages were transmitted to consumers through linear media, including radios and televisions. However, contemporary communication has demonstrated shifts in this model. Modern communication suggests that its purpose is not imparting messages but rather connecting communities through shared beliefs. Under this model, consumers connect through commonalities and interests. Content is shared through interactive delivery vehicles such as the Internet and social media.

This mass communications shift has created a unique culture, which has agitated norms and established institutions. Moreover, technology has greatly accelerated the speed of change. The impacts of these shifts have reverberated within several capacities including industry, academia, and research.

The industry of mass communications has several professions, including advertising, public relations, journalism, and publishing. As a result of triadic convergence many of these industries have been drawn together as the boundaries separating them have begun to disintegrate. This has contributed to hysteria and confusion for professionals as well as educators. Intermingling blurs lines between professional roles and responsibilities. Moreover, industries are facing unprecedented competition for the attention of customers. Due to expanded roles and pressure, professionals have been compelled to increase their skills and knowledge to remain relevant and employed. Furthermore, mass communications educators face severe challenges as they attempt to educate and prepare students to enter mass communications industries that are ever changing (McEachern, 2012).

Another vital element to consider is research. At the core of crisis and instability are discrepancies between theory and fact (Kuhn, 1964). Gaps in research and knowledge lead to naive interpretations and inaccurate methods and practice. Not only will research fill these voids, it will also contribute to problem solving and the end of crisis within advertising. Furthermore, scholarship creates opportunities for increased awareness, inclusion, and discourse. Research assists the reconciliation of perception and reality.

Due to its nature and potential evolution, triadic convergence may cause shifts among many of the key systems, processes, and components within this community. However, within contemporary intermingling of media, technology, and culture there are key factors that will assist in navigation.

Creativity is a driving force in navigation. Creativity is desperately needed to sustain productivity while problem solving ensues. Creativity is also a vital asset to researchers and advertising professionals seeking to uncover the norms, patterns, and solutions associated with triadic convergence. Formidable methods, innovation, and strategies are required until a consensus is reached and a new paradigm is firmly in place. Moreover, advertising professionals and students are compelled to demonstrate fresh insights and creative ingenuity to remain competitive and employed.

The disintegration of the established paradigm has led to varied creative approaches to advertising as professionals attempt to engage audiences. In some instances, creative teams are now instructed to behave more like improvisational actors. Creative teams are encouraged to focus on "story building" and "storytelling" in order to remain nimble in response to real time and migratory behavior. Some advertising professionals have had to become adept at product development, brand utility, and functionality (Sacks, 2010).

In addition to creativity, high levels of self-efficacy are of critical importance. There is no specific delineation for the duration of crisis, therefore, bold resolve is crucial to success. Those who persist in problem solving must be confident of mastery, regardless of outcome. Throughout the quest to achieve consensus, there may be numerous cycles of trial and error. Self-efficacy is the strength of an individual's belief in his or her ability to accomplish a goal. Those with high levels of self-efficacy can effectively organize and execute the actions required to endure and achieve (Cloninger, 2013). In addition to nurturing individuals with high self-efficacy, environments must be created that will cultivate it. Thus, creating safe spaces for failure is critical to overall success.

Key factors in the interaction between technology, media, and culture serve as the impetus for the coalescence that generates triadic convergence. Such factors include the following:

1. Advances in technology drive economic processes of cross-ownership.
2. Advances in technology have a direct impact on media.
3. Characteristics of media ownership are largely determined by characteristics of technology.
4. Whether or not media ownership is typified by monopoly or competition is dictated by the capabilities of the digital environment (de Sola Pool).

Technology and culture have a causal relationship in which the cause-and-effect are sometimes reciprocal and not fixated.

Themes Associated with Triadic Convergence

The remaining discussion will explore dominant points of interest related to convergence. In an attempt to understand the contemporary advertising landscape, one must keep in mind that advertising is currently highly characterized by monopoly. In contrast, modern technology is highly characterized by communal tendencies and interactivity. Thus, the potential for enormous conflicts in advertising is created. Recent mergers have led to an enormous concentration of power and wealth within industry (Deuze, 2010). Media conglomerates command entire areas of entertainment. In turn, technologies have lowered production costs, expanding the range of available delivery channels and enabling consumers to archive, circulate, and interact with media in an unprecedented fashion (Jenkins, 2006). Examples of these new means of interaction include social media platforms, which are explored in the following sections.

Social Media

Social media, a dominant symbol of the magnitude of triadic convergence, is a relatively recent and rapidly evolving platform of media. In the United States alone, social media reaches nearly 80% of active U.S. Internet users and represents the majority of Americans' time online. As a global phenomenon, social media reaches more than 75% of active Internet users. Thus, today's technological landscape includes the pervasiveness of social media and mobile devices, which simultaneously facilitate isolation and interactivity among individuals, communities, and organizations (Sagan & Leighton, 2010).

The elusive nature of social media often makes it difficult to ascertain its true meaning. Generally, social media is considered to be a group of online networks that facilitate participation, openness, conversation, community, and connectivity. Additionally, it has gained enormous momentum through advances in technology, namely increased broadband access and mobile devices. Assuming a variety of forms, social media can range from Internet forums and blogs to social networking, virtual game worlds, and podcasts. Associated technologies include, but are not limited to, crowd sourcing, open forums, files sharing, and postings (Slater, 2002). Also, advances in web 2.0 allow many social media platforms to now include visual components for communicating with pictures, videos, and audio in real time.

Theoretical frameworks associated with the examination of social media delve into its constructs as both a media and social platform. For example, social presence theory (Short, Williams, & Christie, 1976) states that media differ in the degree of social presence, which is defined as the acoustic, visual, and physical contact that can be achieved through media between two communication partners. Social presence is influenced by the intimacy and immediacy of the medium. Accordingly, social presence theory posits that the higher the social presence, the larger the social influence that the communication partners have on each other's behavior. This theory has become deeply entrenched and an influential player in the world of advertising and beyond.

As such, the examination of social media becomes increasingly complex. The significance of social media expands exponentially with advances in technology. To illustrate, the telephone was once solely a hard-wired home and office communication device. However, advances in satellite and broadband technologies have fostered the ability to create cell phones, which have quickly become one of the most versatile forms of media. In addition to conversation, today cell phones provide access to music, movies, photographs, video games, commerce, text messaging, instant messaging, and the Internet. Thus, cell phones and mobile devices are now linked to social media and social networks. Consequently, new opportunities and strategies have developed within advertising to complement social media. For years, advertising has been struggling to effectively cope with a diminished mass audience. Social media holds the promise of audience restoration. Nonetheless, a need for increased understanding remains. Audiences of social media tend to be extremely polarized and subject to fragmentation according to the interests of specific communities.

Twitter

Twitter is representative of another dominant point of interest as it relates to triadic convergence. Although social media is overall promising, Twitter is of particular interest due to its flexible, mutative, and influential nature. Twitter is also deeply connected with professional sports, which is a catalyst to connect to large mass audiences. Twitter is a point of intersection within the intermingling of media, technology, and culture. As a real-time information network, Twitter serves as an example of modern global communities facilitated through technology, featuring a constant stream of public conversation. Users are connected to their customized interests through microblogging and social network services. Tweets are used to post instant personal updates, entertainment, industry information, news, politics, and, most recently, pictures and videos through Instagram. Users participate in the conversation through accruing followers, posting information, and sending direct messages (Berinito, 2010). Launched just three years ago as a prototype for interoffice communications, today Twitter has achieved exponential success boasting well over 3 million users, records of nearly 7,000 tweets per second, 460,000 new accounts per day, and simultaneous broadcasts in 43 foreign languages. On an average day, 140 million tweets reverberate within the blogosphere (Rao, 2011).

Moreover, Twitter features rich capabilities to measure data and is gaining strides in its ability to monetize. Advertising has already utilized Twitter to successfully engage audiences and is expected to benefit even further in accordance with Twitter's technological advances. A great deal of consumer research is derived from Twitter user comments, trends, and feedback.

Twitter offers more than 70,000 applications that complete tasks ranging from providing respondent location information to finding discussion groups and determining extent of influence (Rao). One of Twitter's most impressive areas of applications lies within the field of data visualization. Data visualization assists in providing insights for sparse and complex data sets through communicating key features in an intuitive capacity (Friendly, 2009).

Twitter also has the capacity to provide a rich historical database. Approximately 90% of all tweets are public (Rao). Consequently, millions of users worldwide are providing history in real time. As understanding of Twitter develops and Twitter users increase, this information will be an incredibly valuable resource for advertising. In April 2010 Twitter donated access to its entire archive of public tweets to the Library of Congress for preservation and research (Raymond, 2010).

Several companies have already devised strategic methods to utilize Twitter. Advertising has developed ways to make brands and products worthy of conversation among Twitter users. There are noteworthy cases that serve as models throughout advertising, including Carphone Warehouse and Zappos.

The Carphone Warehouse (CPW) is Europe's leading independent retailer of mobile phones and services. The company has more than 2,400 stores in nine countries (Petough, 2010). It recently learned that customers were posting comments on Twitter regarding their dissatisfaction. The owner noted that the company could not control whether customers complained or posted the content of the complaints. However, it realized customer complaints are a positive driving force for making needed changes within the company. CPW made the decision to publicly acknowledge customer complaints and issue a formal apology using Twitter as part of its customer service technology. CPW's approach to Twitter use started with two accounts, @guy1067 and @carphoneware, and eventually added more (Petough). After conducting its own private research, the company developed additional critical insights regarding its consumers.

Twitter has also been successfully used by Zappos, one of the largest online shoe stores in the world. During the early stages of its history the company harnessed the power of social media to aggressively interact with their customers. There are currently 450 Zappos employees with Twitter accounts and they have a collective following of more than 1 million customers (Hendry, 2008). Zappos actively engages in consumer initiatives, including contests in which they ask their followers to help them rewrite confirmation emails and additional loyalty verbiage. Zappos's use of Twitter serves multiple business goals at once. It provides valuable feedback to help improve customer experience. It also affords a medium for the salesforce to generate new customers through direct outreach. Moreover, it reinforces the branding of Zappos as a customer service–related business that cares about connecting with its core consumer (Hendry).

Additionally, Twitter will continue to increase in value as it becomes further interwoven within contemporary culture. Advertisers have the ability to influence, persuade, and engage users as a result of Twitter's enormous following among popular culture influencers, including celebrities and sports figures. Twitter also implemented methods to allow photographs and videos to automatically display in users' streams. As a result, the Twitter stream has the capacity to be populated by rich media. Advertisers can now publish large display advertisements and share the advertisements with users. Moreover, Twitter only charges advertisers when a user engages with a promoted tweet, which includes behaviors of "favoriting" it or retweeting it. With advertisements that feature large rich media, engagement among Twitter users may be increased (Delo, 2013).

Twitter's aesthetic adjustments have helped to bolster its capacity to influence. As previously mentioned, social presence theory posits that the higher the social presence, the larger the social influence the communication partners have on each other's behavior. To illustrate, Twitter has become a vital component of modern popular culture, particularly sports, in just a short time. This is of great significance because professional sports offer advertisers the opportunity to connect with one of the few mass audiences that remains (Oates, 2009). Increased fragmentation and migratory consumer behavior have made it difficult for advertisers to connect with mass markets.

Research affirms Twitter's dominant presence in sports and its ability to create a powerful mix of influencers interacting with audiences. This creates the unique opportunity for advertisers to engage and connect with an influential mass audience. Twitter is used to provide commentary and opinion by players, fans, and media professionals. Twitter provides a platform for a diverse range of high-profile athletes, including Lance Armstrong (cycling), Serena Williams (tennis), Usain Bolt (track and field), Lote Tuqiri (rugby), and Shaquille O'Neal (basketball), to engage in instantaneous, unpredictable, and occasionally controversial communication and promotion with followers (Hutchins, 2011). Every NBA, MLB, NHL, and NFL franchise has a substantial Twitter presence, as well as roughly half of all professional athletes. Retired NBA player O'Neal has more than 4 million followers, more than the combined total of the daily readership of the *New York Times* and the entire viewing audience of the 11:00 p.m. televised broadcast of *Sports Center* (Wertheim, 2011).

Super Bowl 2013 shattered previous social media records, with reports of 30.6 million user comments. Astoundingly, of the 30.6 million comments, 24.1 million posts were generated via Twitter. Moreover, Twitter reported that nearly half of the national television spots aired during Super Bowl 2013 included a hashtag, a word or phrase preceded by a number symbol that's utilized to organize subjects on Twitter. It is estimated that Twitter generated advertising revenue of $545.2 million in 2013, an increase of 89% over 2012. International projections for 2014 estimate that advertising revenue from Twitter will soar to $807.5 million, a 48% increase over 2013 (Ortutay, 2013).

Rapid Progression of Technology

The exponential growth and influence of Twitter affirms advertisers' need for constant awareness of the rapid progression of technology. In 2010 the United Nations confirmed that the number of worldwide Internet users had surpassed 2 billion, with nearly 3 million users in North America alone. Computers are now linked with telephones, satellites, and other forms of mass media to create networks of global communication, thus, facilitating a central electrical environment in which information can be stored, manipulated, distributed, and presented (Alkalimat, 2004). Spending nearly 700 hours online each year, American consumers rely on the Internet for the completion of varied essential tasks, ranging from business and entertainment to education and leisure (Nielsen, 2011).

The rapid pace at which technology proliferates the business and interaction of society can often elicit frenzy; however, the contemporary nature of technology is highly characterized by deregulation. Although this presents a threat to order and can be construed as frenzy, the deregulation of technology empowers audiences like never before. Technology enthusiasts advocate the increased command of technology through consumer uptake; that is, they insist that technologies are simply delivery systems—the tools used to access media content. Media are cultural systems that persist as dynamic layers within society and lifestyle. Consequently, to address the impact of convergence from merely a technological perspective would be an error. History has revealed that delivery technologies often become obsolete and are frequently replaced, while media continue to evolve. Traditional media, including print, radio, television, and cinema, are forced to coexist with emergent media, including social media, social television, and mobile devices (Jenkins, 2006).

Currently, traditional media is not being displaced but rather shifting as a result of the progression of technology. Some strategic advertising models implement a total approach in which various delivery technologies, which include traditional and emergent media, are used to communicate messages to audiences. According to a recent Interactive Advertising Bureau survey of brand marketers, mobile budgets surged 142% between 2011 and 2013. Understanding this new market, as well as its users, may create opportunities for increased engagement, brand lift, product purchasing, monetization, and reliable analytics (Gallagher, 2013).

Participatory Culture

As technology and media evolve, culture will as well. In fact, triadic convergence reflects many of the distinct transformations involving how contemporary media is produced and consumed. Producers and consumers of media interact within a seemingly contradictory, yet dynamic, reciprocal relationship. States Jenkins:

> If old consumers were assumed to be passive, then new consumers are active. If old consumers were predictable and stayed where you told them, then new consumers are migratory, showing a declining loyalty to networks or media. If old consumers were

isolated individuals, then new consumers are more socially connected. If the work of media consumers was once silent and invisible, then new consumers are now noisy and public. (pp. 18–19)

Jenkins explains that the construct of participatory culture is to assist in understanding contemporary media environments characterized by convergence. Participatory culture involves communities in which members are actively involved in the creation and circulation of media content. It is derived from the complex relationship between the consumption and production of media. As demonstrated through recent industry trends, advertising professionals frequently invite consumers to participate in contests, testimonials, and discussion forums, among other events, regarding their favorite brands (Jenkins). Advertising professionals use this information to develop insights upon which to build advertising campaigns and to leverage communication tactics with audiences. Conversely, audiences utilize blogs, social media, trends in subcultures, and so forth, to offer relevant media content to advertisers. Such content is not typically a part of the mainstream media and, therefore, provides increased relevancy and opportunities for greater successes for brands and advertising campaigns (Christodoulides, Jevons, & Bonhomme, 2012).

In a similar capacity, shifts in America's demographic and generational landscape have contributed to increased cultural activity. Culture is an evolving set of shared beliefs, values, attitudes, and logical processes that provide cognitive maps for people within a given societal group to perceive, think, reason, act, react, and interact (Tung, 1996, p. 244). According to recent U.S. census data, 1 in every 3 Americans is a person of color; adults over the age of 65 constitute the fastest growing population segment; and specialty markets driven by female buying power comprise 84% of the U.S. population (Lafayette, 2011). Moreover, America's largest population segments are inherently participatory. Baby boomers and millennials are diverse and highly participatory, characterized by communal tendencies and deep engagement with media (Howe & Strauss, 2000). Therefore, understanding these audiences will be critical to the success of advertising. Each generation will be explored respectively, particularly as they are relevant to the contemporary triadic convergence experience.

Baby Boomers

Like culture, American baby boomers are continuously evolving as the years progress. In fact, boomer development is an ongoing evolutionary process that involves changes in priorities and values at individual, generational, and societal levels (Ergi & Ralston, 2004). The boomer generation is typically considered to have been born from 1946 to 1964. They are comprised of more than 79 million individuals. Heterogeneity is one of this generation's most consistent and defining attributes. Boomers are educated, ethnically diverse, liberal, conservative, politically active, demanding, and technologically savvy. Their varied participatory experiences have continuously propelled them to challenge the status quo. Members of this generation have questioned everything from fashion and education to race relations, sexual mores, and gender roles. Boomers have helped shape and redefine nearly every aspect of life (Glass, 2007).

Undoubtedly, boomers will redefine the aging process in America as they are healthier, more vital, and have longer life expectancies than previous generations. In 2011 the first boomer cohort reached 65. For the next 17 years, it is projected that more than 10,000 people will celebrate their 65th birthday each day (Kennedy, 2010). These maturing markets control more than three-fourths of America's wealth and outspend other generations by approximately $400 billion each year on consumer goods and services (Lockwood, 2014). Boomers are the highest earners, with a median household income of $54,170, 55% greater than post-boomers and 61% more than pre-boomers. Additionally, boomers are the most influential investing group, with 40% of the U.S. population over age 50 controlling 75% of financial assets and 50% of all consumer spending (Kennedy & Mancini, 2006).

Boomers grew up during a period of unrivaled prosperity, advertising dominance, and affluence following World War II. Members of this generation were active in radical change, namely the civil rights movement, protests against the Vietnam War, the women's rights movement, and the communications technology era (Howe & Strauss, 2000). Boomers came of age during the dramatic upheaval of American family life in the 1970s, which was characterized by nonmarital childbearing, unprecedented divorce rates, cohabitation, and delayed and dissolved marriages. Research indicates that such diverse experiences have impacted boomers, such that, for example, one in three boomers is unmarried and is an avid online dater (Cherlin, 2010).

Boomers have demonstrated a strong work ethic and high job involvement, which has led to career success (Howe & Strauss). Accordingly, boomers were hit hard by the economic turmoil of the Internet dot-com investment bust, the real estate crisis, and the recent national recession. Boomers reject the notion of leisurely retirement and plan to work as long as they are physically and mentally able to do so. These characteristics are promising for advertising as such behavior indicates not only a skilled and flexible workforce, but also a substantial amount of discretionary income (Kennedy & Mancini).

Millennials

Millennials, the advertising industry's primary audience, is America's most diverse cohort to date. Research reveals that the millennial population is most representative of the effects of convergence (Pasek, Kenski, Romer, & Jamieson, 2006). As this generation matures, sectors of life, industry, and commerce will be revolutionized in accordance with their distinct characteristics (Howe & Strauss).

Although the exact dates of origin and conclusion tend to vary slightly, millennials typically were born between 1982 and 2002 (Howe & Strauss). Among millennials ages 13 to 29, 18.5% are Hispanic; 14.2% are black; 4.3% are Asian; 3.2% are mixed race or other; and 59.8% are Caucasian (Keeter, 2010). Millennials have been noted for their high levels of tolerance and desire for self-expression and individuality. Additionally, millennials are on course to become the most educated generation in American history. Hit the hardest by the recent economic crisis, the International Labor Organization reports that 4.5 million millennials are unemployed. Consequently, some millennials have opted to further their education with secondary degrees. Census data reveal that millennials are slated to become America's largest population, with numbers surpassing those of baby boomers and other previous generations. Totaling 86 million, recent data bolster this estimate to 100 million with the inclusion of immigrant populations (Orrell, 2009).

Moreover, millennials wield $170 billion in purchasing power. Known as digital natives, millennials are the first generation in American history that have grown up totally immersed in a world of digital technology. Consequently, this generation is generally described as heavy users of, and extraordinarily familiar with, communications devices, mass media, and digital technology. They characterize this usage as the primary defining characteristic of their generation (Keeter, 2010) and regard behaviors like tweeting and texting, along with the use of other social media, as normal and integral components of daily life. According to a 2011 study conducted by Cisco Systems, millennials view the Internet as just as important as air, water, food, and shelter (Stricker, 2011).

It is important for advertisers to know the needs and wants of the millennial and baby boomer generations, as they are currently the largest generational cohorts within participatory consumer audience segments. The propensity to create successful advertising strategies may be achieved by connecting consumer insights with knowledge of the convergence.

These findings are not intended to provide a detailed characterization but rather a synthesis of major themes as they relate to the contemporary crisis within advertising. There are numerous stakeholders within the new media ecosystem. Information and awareness are critical. Triadic convergence has created ways to propel a rich and diverse agenda into the public sphere through advertising.

· 8 ·
MOVING FORWARD

Advertising in Crisis

Advertising is in a state of crisis. History, theory, and context affirm its presence. Collective productivity within the discipline has ground to a halt. Stagnant earnings, increased costs, disintegrating relationships, futile strategies, fragmentation, panic, and massive layoffs characterize much of the industry. Analysts have predicted more downsizing and restructuring, leaving many advertising agencies to contemplate their future amid major cutbacks. Echoing the downsizing trend, multinational conglomerates such as Unilever and Procter & Gamble have announced massive job layoffs. Moreover, major brands, including Coke, Pepsi, and Avon, have reported impending advertising reductions (Neff, 2013).

In the face of this crisis, many advertising agencies have been forced to restructure or dissolve (Tungate, 2007). Plummeting advertising budgets have reached declines of nearly 54% and are projected to continue in an indefinite downward spiral (Lee, Taylor, & Chung, 2011). Several premiere shops, small agencies, and creative boutiques have collapsed. Adding to this confusion, recent cutbacks and lean practices have led to overlapping professional responsibilities and blurred boundaries.

In addition to trends in downsizing, the monopolizing prescience of conglomerations in advertising continues to persist. Global conglomerates, including the WPP Group, Publicis Groupe, the Omnicom Group, and Interpublic Group (IPG), wield significant holdings and influence throughout advertising and global communications (Inukonda & Pereira, 2010). The landscape of advertising is expected to continue in this direction, which may exacerbate a number of pressing issues, including data analytics, market maturation, creative growth, and personnel disruption (Bruell, 2013).

Other indications of crisis include transformative patterns in production and consumption, which have altered productivity. Examples of these patterns include drastic shifts in traditional media behavior. Target audiences have evolved into savvy migratory communities, leaving advertising professionals struggling to determine effective methods for engagement and connectivity. Additionally, rapid proliferation of new communication technologies has reconfigured markets at an exhaustive pace. Conditions have been exasperated by difficulties surrounding measurement, monetization, and sustained revenue. Advertising is struggling to exist within this contemporary environment.

Although this state of crisis has been accompanied by calamity and turmoil, it is also the precursor to extraordinary transformation. It is only through crisis that the path to a paradigm shift will emerge.

Paradigm Shift

A paradigm shift is a change in the governing principles of advertising. A complete paradigm shift will occur when advertising collectively adopts a new paradigm in order to restore productivity. The prescience of crisis in advertising affirms the impending paradigm shift. Evidence is further verified through key characteristics of disintegration and patterns in history. Key characteristics that verify the impending paradigm shift are complexity and blurring. History provides a valuable narrative that traces ideology, systems of regulatory power, and patterns of conflict and resolution in advertising. Within its unfolding, the development of advertising reveals outlines of the past and projections for the future.

Information such as this has incredible value, as a paradigm shift presents the opportunity to align advertising with the future. Paradigm shifts are accompanied by a phenomenon that drastically altered the world and individuals' perceptions of reality. Thus, a paradigm shift is a malleable juncture during which the future can be shaped. Knowledge of the past and present will assist in the ability to avoid cyclical patterns of marginalization and disruption while creating optimal solutions.

Characteristics of Disintegration

Both crisis and paradigm shifts begin with the blurring of practice and ideology, which inevitably leads to massive disintegration. Blurring loosens the rules that govern a community. Since its inception, blurring has been an intrinsic attribute of advertising. As the field transformed into a profession, blurring became a part of its foundation. The 1920s distinctly mark the beginning of the blurring of advertising and popular culture. In the ensuing decades, advertising has worked vehemently to blur the lines between product and culture in order to create links from products to lifestyle, trends, and socially significant behavior (Danesi, 2012). The blurring of popular culture and advertising contributed to the complex intermingling of media, technology, and culture. Blurring plays a significant role in the new media ecosystem. As content flows across various platforms, little to no distinctions are made among voice, video, images, and text. Content has become blurred data. Moreover, boundaries between consumers and advertising professionals have been blurred through the popularity of user-generated content. The proliferation of these entities, combined with the far-reaching influence of advertising, exacerbates complexity.

Like blurring, complexity is a necessary precursor and major characteristic of an impending paradigm shift. Complexity signals a state of crisis. In fact, complexity is what distinguishes crisis from an anomaly of normal occurrence. Throughout its early beginnings, new and unaccounted for anomalies arose within advertising. In response to anomalies, new advertising theories, models, and practices developed. However, advertising anomalies that are characterized by complexity cannot be resolved using traditional methods. There are attempts to isolate the anomaly, thus, ensuring the productivity of the discipline. However, complex anomalies refute isolation. Advertising issues that are suppressed in one area resurface in another. The blurring that characterizes complexity will bolster its continuous intermingling force. In this scenario, crisis erupts and continues indefinitely.

Disintegration associated with complexity is intensified by advertising's interdisciplinary nature. Complexity is exacerbated as one particular paradigm is used to govern the practices of multiple interwoven disciplines. Advertising's core interwoven disciplines, including business, art, and psychology, are uniquely focused on varied goals of research and practice. Thus, when crisis erupts it becomes increasingly difficult for a sole paradigm to determine several overlapping traditions while sustaining productivity.

History

History offers a unique lens to observe the development of advertising. Not only does it provide the opportunity to trace the development of blurring and complexity within advertising, but it also offers critical insights. Historical context reveals the dominant power structures that are closely aligned with the established paradigm in advertising. In addition to historical and societal context, power structures have a direct impact on patterns surrounding conflict and resolution. Furthermore, the observation of power structures cultivates opportunities to create awareness surrounding the potential to redistribute power.

Overarching, history has revealed a rhythmic pattern in which the pendulum of power has swayed back and forth for well over a century within advertising. However, throughout its development distinct instances of blurring and complexity cultivated an environment for crisis and the impending paradigm shift. An accumulation of these elements became noticeable toward the close of the 1980s.

In the 1980s major power disruptions exploded and continued well into the early 1990s. Acute delineations of initial impact are difficult to identify due to widespread blurring endemic to advertising. As an interdisciplinary profession, the pendulum of power has continuously swung among the three core disciplines within advertising, namely business, science, and creative. However, the flow of this rhythm was disrupted by the duration of the stronghold of business conglomeration within advertising. Beginning with the 1970s, the influence and expansion of business conglomeration in advertising skyrocketed and resulted in intense monopoly. This was in sharp contrast with the short dominance of creativity and inclusion experienced during the 1960s in what was known as the creative revolution in advertising.

Prior to this occurrence, advertising developed a pattern of normalcy in which the pendulum of power swung between interdisciplinary disciplines within advertising. Moreover, the established paradigm of advertising affirmed that successful outcomes are achieved through interdisciplinary collaboration and the ability to harness technology and modernity. History reveals that in the 1970s, power shifted toward business in advertising for a sustained period of time. This change resulted in an excess of corporate influence and power, which was not indicative of the longitudinal norms established within the discipline.

Thus, when triadic convergence collided with internal structures of advertising in the 1980s, widespread calamity erupted. Advances in electronification fueled triadic convergence. This was made evident largely through the impact of technology produced by Adobe. The democratization of technological information fostered by Adobe not only alleviated tensions surrounding power, but it also shifted inherent structural systems within advertising. Power relationships were upset at an unprecedented level.

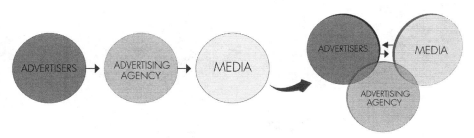

Figure 8.1. Internal Disruption Model.

Advertising experienced numerous power disruptions within its internal structures. Power shifted within advertising agencies and from an intense stronghold among business executives and clients to advertising creative practitioners. Power also shifted among vendor relationships. For example, advances in electronification created hardware and software technologies that diversified business affiliations and production alliances. Power also shifted from conglomerations to independent agencies, consultants, and boutiques. In some instances, independent agencies, consultants, and boutiques carved a niche by offering specialties to clients, including those in media, culture, and technology. Increased competition created rivalries within advertising as agencies began to lose their footing with advertisers.

In stark contrast with its early beginnings, many advertising agencies were forced out of their established role and relationship with advertisers. Advertisers began to work directly with media and consultants instead of through advertising agencies. This disruption of power was extremely significant because it involved shifts in knowledge and information, as well as shifts in economic structure. These shifts resulted in simultaneous calamity and empowerment, which is a contradictory occurrence of crisis.

While still reeling from calamity experienced within its inherent structure, advertising collided with triadic convergence yet again just a short time later. This collision upset external structural systems in advertising. In the

1990s modern advances in electronification were extended to the masses. Triadic convergence was manifested through the commercialization of the Internet, which contributed to the rise of global technology around 1992. Just as technology had empowered advertising creative professionals, consumers also were empowered.

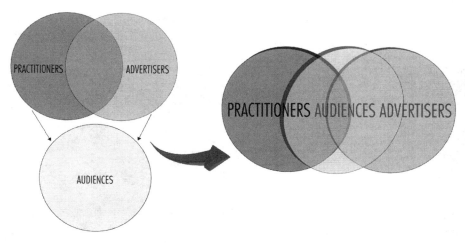

Figure 8.2. External Disruption Model.

Further disruptions have been caused by increased access and information technology, which has led to shifts in power relationships between consumers and advertisers. These external disruptions compounded the calamity that advertising experienced internally. Furthermore, instabilities within advertising's external environment spread rapidly worldwide. Domestic and international markets became muddied and experienced confusion due to increased intermingling.

Empowerment contributed to the further disintegration of the established paradigm of advertising. For example, empowered consumers were no longer as easy to persuade or influence, therefore, persuasion tactics proved insufficient. Advertising had to create new ways to engage consumers in a short time while immersed in calamity and chaos. The stronghold of business led to a suppression of interdisciplinary efforts, which historically facilitated advertising's ability to create new ways to persuade consumers. Moreover, the widespread proliferation of factions inhibited opportunities for collaboration. Additionally, advertising was no longer able to successfully rely upon its ability to harness technology. New media and technology were in flux, leaving new technologies unreliable and unable to sustain the weight of the institution of advertising.

Triadic convergence facilitated increased comingling and empowerment. As a result, the creation of advertising content began to blur. Consumers and practitioners became involved in content creation in ways that upset the organization of advertising. In the 1980s advances in electronification and technology empowered practitioners to create and produce at levels similar to that of professional vendors. This resulted in a major upset of roles, relationships, and responsibilities. In the 1990s further advances in electronification and technology allowed consumers to create at a level similar to that of advertising practitioners. This trend resulted in widespread chaos.

Modern consumers have indicated that they no longer consider themselves as merely consumers and dislike the conceptualization of target audiences used within advertising. Instead, trends indicate that modern consumers consider themselves as distinct communities involved in the co-creation of content. Contemporary advertising is comprised of complex and contradictory conundrums that have stymied members of the profession.

Triadic Convergence: Support & Criticism

Triadic convergence is a radical force that has revolutionized virtually every facet of life and industry. It is the revolutionary phenomenon that has greatly contributed to the impending paradigm shift in advertising. As a source of modern controversy it has accumulated both support and criticism. Nonetheless, its impact is most readily apparent in its potential for extraordinary achievements. Some of these achievements include advancements in communication technologies that allow advertisers and users to engage in collaborative and interactive communication with rich media. Through increased interactive communication, some brands have been able to understand consumer needs and make stronger connections.

Triadic convergence has contributed to the decisions of some brands to deliberately work toward fundamental social change, while meeting the needs of advertisers and consumers. During the conference of the Committee Encouraging Corporate Philanthropy (CECP), 55 CEOs from many of the world's largest companies recognized the need to "elevate and prioritize the social contract" and to "connect their strengths to the acute needs of the communities in which they do business" (Henderson, 2009).

Triadic convergence has also contributed to advertisers' shifts toward aligning with more inclusive cultural markets. Brands have noted that multicultural consumers are now perceived as the new mainstream rather than a niche market. In a manner that also typifies triadic convergence, the advertising conglomerate Publicis Groupe is slated to undertake the new campaign for the Toyota Motor Company's strategic alignment with multicultural consumers. These efforts will be led by Publicis Groupe's Saatchi & Saatchi agency, along with three sibling Publicis Groupe shops, Conill advertising (Hispanic), Burrell Communications (African American), and Zenith (broadcast and out-of-home media buying). interTrend Communications, which specializes in Asian American communications, will also be part of the venture (McCarthy & Wentz, 2013).

In a similar capacity, triadic convergence has contributed to diversity and inclusion efforts within hiring and retention practices in advertising. The industry has long since been accused of lagging behind cultural progress as well as reflecting atrocities within society. Thus, advertising has received criticism for racist, sexist, ageist, and classist practices. In many ways triadic convergence has become an equalizer for those wishing to enter advertising or maintain long-term success. The need to connect with shifting demographics is compelling advertising to diversify at comprehensive levels, which range from entry to executive. Moreover, advances in media and technology have facilitated autodidacticism and independent ventures. Diverse members within the advertising community may also contribute to innovation and overall success.

Triadic convergence has also contributed to achievements acquired through the results of widespread mergers, acquisitions, and collaborations and partnerships within media, technology, and entertainment sectors. Converged sectors have led to increased engagement among participatory audiences through the development of content influenced by real-time technologies, social data, and cultural insights.

In addition, triadic convergence has contributed to consumer empowerment. Consumers have been afforded methods and devices to interact directly with advertisers. They have also been empowered through opportunities created by fan-voted endings and live audience voting systems. In turn, new and inventive research methods and analytics, as well as mobile devices and applications have also been incorporated within social television platforms through audiences' simultaneous usage of multiple screen devices. Adaptive campaigns, which feature relevant cutting-edge content, including multiple screen mashups, promotions, bundles, and consumer loyalty incentives, have been created. In sharp contrast with previous times, modern handheld mobile devices are powerful computers that are not just cell phones or tablets but also are cameras, MP3 players, radios, gaming devices, electronic organizers, the Internet, storage devices, instant messaging systems, and television. Due to triadic convergence, one can conceptualize seamless communications among device, media, technology, network, and industry. Accordingly, advertisers are beginning to make progress as they develop strategies and metrics that help them understand how participatory cultures engage with content, media, and technology (Redniss, 2013).

Despite the potential for extraordinary achievement, there are other conditions that indicate a continuing state of crisis. These conditions, which are characterized by complexity, competing solutions, and declines in productivity, contribute to diverse and contradictory views of triadic convergence. Thus, the future vision for advertising reveals multiple perspectives. These perspectives offer competing solutions from various sectors and factions within advertising. While some have sought to mitigate triadic convergence, others have chosen to leverage triadic convergence.

Some corporate advertisers have leveraged aspects of triadic convergence to exploit advantages associated with media conglomeration. Clients have reduced costs by partnering directly with media or advertising conglomerates rather than with established advertising agencies. In addition, triadic convergence has resulted in numerous ways to reach consumers directly and to attain their loyalty. Triadic convergence–based advertising strategies can be a key factor in shaping consumer behavior during this time of extensive fragmentation. Consumers who demand more customized responses and interactive experience interests tend to advocate convergence-based strategies (Jenkins, 2006).

Companies with a vested interest in technology avidly support triadic convergence as a way to emphasize and demand creativity and modernity, thus reinforcing the established paradigm of persuasion and techniques involving technology. Formidable advertising agencies propose strategies that emphasize more aggressive forms of integrated marketing communications (IMC). Moriarty, Mitchell, and Wells (2012) indicate that such strategies demonstrate an effective ability to proliferate audiences and sell products by creating an emotional bond with consumers. Advocates of these tactics contend that the restoration of the discipline lies with brand utility (the act of providing something useful, relevant, or entertaining that deeply embeds itself within everyday life), communal activities, culture, and experience (Bernardin & Kemp-Robertson, 2008).

Communal activities are emphasized through experiences with brands and products. Branding and consumer loyalty are reinforced through added value and frequency. Interactive components of these solutions encourage people to go online to share their experiences. This stage of the process provides the opportunity to measure and capture behavioral data. These solutions are based within the new paradigm coined "human kind." This approach mandates that consumers be treated as people. Humanity and relationships, rather than brands or products, are the central theme and guiding factor (Bernardin & Kemp-Robertson).

In a similar capacity, marketing experts agree that the future success of advertising lies within its abilities to harness culture. However, critics of integrated marketing communications suggest that the answers will not be found there. Rather, Arnould and Thompson (2005) assert that the identification of human passion points and universal truths will be critical to unifying culture. Proponents affirm the use of consumer culture theory (CCT), which provides insights to cultural meanings, sociohistorical influences, and social dynamics that shape consumer experiences and identities in the context of lifestyle and society. Moreover, supports of CCT indicate that the use of this theory may mitigate the rampant use of stereotypes and cliché imagery within advertising. CCT research stresses that consumers are not unified, monolithic, or transparently rational (Belk, Guliz, & Soren, 2003).

Opponents of this approach contend that a renewed emphasis on strategic and well-disciplined creative is the most appropriate solution. They contend that more research must be conducted on creativity. Stewart, Cheng, and Wan (2008) affirm that a balance struck between creative spontaneity and discipline is the key to achieving sustained success. Cognitive psychologists assert that theoretical and empirical findings must serve as the foundations upon which to build effective creative advertising. Proponents suggest an overhaul of the creative process and creative briefs. They cite recent advances in cognitive processing and neuroscience as fertile ground upon which to build the future of advertising (Lorigo, Haridasan, Brynjarsdottir, Xia, Joachims, Gay et al., 2008).

In an era characterized by an impending paradigm shift, numerous proposed solutions seem valid, which contributes to chaos and confusion. Increased awareness and active problem solving is needed to cultivate optimal conditions. Triadic convergence has the potential to propel advertising toward unforeseen achievement. More knowledge regarding this phenomenon would undoubtedly lead to advantageous outcomes.

Moving Forward

The impact of triadic convergence on advertising is a global concern. The impending paradigm shift, which has resulted from the collision of these dynamic phenomena, will reverberate for centuries. Not only is advertising an institution of enormous influence, but triadic convergence is a powerful force capable of fueling endeavors to stratospheric proportion. A future shaped by the outcome of this intricate collision will impact major areas, which include the economy, mass communications, lifestyle, and technology. Despite its magnitude, awareness is needed.

Although there are a number of concerns that contribute to the need for awareness, this book is particularly oriented with promoting empowerment and other opportunities that accompany change. The most formidable characterization of change and empowerment involve the power disruption in advertising. As a result of the impact of triadic convergence power has been redistributed.

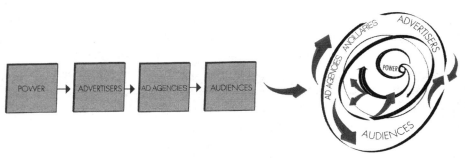

Figure 8.3. Power Disruption Model.

Triadic convergence and modern technology have dispersed power among diverse pools of communication professionals and consumers, leading to their empowerment. This has led to a growing number of factions and competition within advertising. It has also diminished the level of influence advertisers can exert over consumers. Due to triadic convergence, consumers and advertisers have been drawn together in an equalized space. As a result, interactive communication between diverse communities has ensued.

This creates enormous opportunities as advertising is closely aligned with national and global economies, government regulation, and such business practices as mergers and conglomerations. Thus, systems of regulation and power blocs that uphold structural systems within advertising are of significant importance. Of equal importance is critical dialogue that is distinctly concerned with creating and understanding power disruptions and potential redistribution parameters.

An area of particular interest lies within corporate influence in advertising. Conglomerates are at the nexus of economic wealth and power within advertising. As stakeholders, corporate response to increased competition and consumer empowerment may result in tactical attempts to quell the dynamism of triadic convergence. History reveals that advertising has traditionally favored linear communication models in which power was located among advertisers rather than consumers, many of which have now amalgamated through conglomerations.

History has also revealed that phenomena characterized by complexity and competing solutions have been resolved by methods that ensure mass persuasion. Within advertising such resolutions have been characterized by audience segmentation. Segmentation is often favored because it is easier to control consumer behavior that is not intermingled. Although this may provide a seemingly simple or economically profitable outcome, segmentation may lead to instability in the future. Triadic convergence is a powerful force that is difficult to control or suppress. Moreover, triadic convergence spurs communal behavior, interactivity, and increased intermingling. Audience segmentation has even deeper implications surrounding America's changing demographics. As normative culture and lifestyle patterns shift, audience segmentation may have harmful effects, which negates the powerful communal impact triadic convergence has on mass communications.

Power redistribution in advertising is an indicant of the impending paradigm shift, therefore, it is exacerbates the need for awareness. A new paradigm is often invisible as it may emerge prior to the resolution of crisis. During problem solving, it is normal for advertising professionals to attempt to restore productivity expeditiously. Cultivation of a new paradigm may be laborious. Thus, in an infinitesimal state a new paradigm may be shrouded unintentionally. Nonetheless, opportunities to empower consumers and create a more inclusive public sphere may become obscure.

Paradigms may also be hidden by tactical agendas. During a state of crisis, which precipitates a paradigm shift, rhetorical and empirical attacks become rampant. There is a desire to maintain power typically associated with established structural systems and power blocs. Powerful groups maintain their legitimacy through undermining alternatives. History has affirmed that series of knowledges "have been disqualified as non-conceptual knowledges, as insufficiently elaborated knowledges: naive knowledges, hierarchically inferior knowledges, knowledges that are below the required level of erudition or scientificity" (Foucault, Bertani, Fontana, Ewald, & Macey, 2003, p. 7). Advertising must remain vigilant to safeguard against offensive and inaccurate rhetoric that could result in setbacks. Invisibility diminishes the opportunity for inclusive participation, further prompting the need for study in order to promote awareness.

Ascension to a new paradigm is largely motivated by choice, which further intensifies the need for knowledge and awareness. In its fragmented state advertising may be largely unaware or unable to form a collective agenda. Accordingly, the opportunity to shape the future of advertising may be decided by those who have the most wealth, influence, and resources. Due to the far-reaching magnitude of this institution, informed and inclusive decisions are ideal.

Through knowledge and awareness the opportunity to establish a consensus within the discipline and an inclusive agenda can be achieved. Progress toward new horizons for advertising can be informed by a reexamination of current conditions and collective projections for the future.

During this period of reexamination and future projection, there are beacons that may be valuable resources. Exemplary advertising professionals, especially creative practitioners, possess tacit knowledge and are of enormous value during an impending paradigm shift. Additionally, young professionals are of particular importance. Further inquiries surrounding ideation and capitalism offer insights as well.

Within advertising there are exemplary professionals that demonstrate exceptional prowess. These members of the discipline possess tacit knowledge that is vital in the development of insights that will restore productivity (Kuhn, 1964). Tacit knowledge is acquired through practice rather than the acquisition of rules and theory. Neural apparatus translate to experience, judgment, and intuition that are in sharp contrast to theoretical constructs. Thus, problem-solving efforts may be greatly enhanced by the inclusion of firsthand accounts from exemplary practitioners wielding tacit knowledge. Collaboration among scholars, researchers, and advertising professionals may yield results that will reconcile theory and practice.

The ability to understand the impact of triadic convergence on advertising from the perspectives of those who drive its practices is also critical. Literature has not been overly concerned with the perceptions of advertising creative practitioners. However, research suggests that data sets attained from creative practitioners are rich and elaborate (West, Kover, & Caruana, 2008).

Moreover, creative practitioners are often the first to encounter anomalies and engage in problem-solving techniques (Kuhn). Accordingly, knowledge and understanding will expand with the addition of this unique perspective.

Valuable resources may also be found among young advertising professionals. It is beneficial to pay close attention to the tacit knowledge and practices of younger practitioners. Fresh professionals often possess a boldness and nonconventional approach to their work. Those who have been members of a discipline for a short time are often less entrenched in traditional practices and more likely to embrace radical methods and concepts.

Moreover, ideation and creativity are sorely needed in advertising. Incubators for innovation may assist problem solving and diminish poor market conditions. Complexity resulting from saturated markets has continually plagued advertising. History has revealed that economic downturns resulting from market saturation have been a reoccurring theme throughout its development. Advertising began with ideas that translated to products. In turn, these products generated product parities and saturated markets. Over time, advertising implemented techniques such as branding, demand creation, and affiliation to diminish the effects of market saturation. However, after decades of intense consumerism, fragmentation, and saturation, both advertising and the economy are struggling.

Additionally, parallels between crisis in capitalism and crisis in advertising may further provide insights. The global environment is suffering due to mass consumption. Among other things, advertising is suffering due to consumer apathy and migratory behavior. Modern consumers are interested in corporate philanthropy, convenience, and experience more than ever before. The popularity of shared networks, games, mobile devices, and social applications suggest the public's growing desire for services and experience, rather than the purchase of traditional products. Thus, advertising's ability to effectively commodify contemporary conceptualizations of service may provide fresh and insightful solutions.

In addition to insights gained from practitioners and explorations involving ideation and capitalism, the ability to embrace complexity and understand the power of blurring may lead to further insights. Well over a century ago, advertising was created with a specific economic agenda. During its early beginnings, advertising provided information to consumers about products and ensured the organization and stability of a free enterprise system. However, once advertising began to blur with popular culture, new roles, functions, and meanings evolved.

The inextricable relationship between advertising and popular culture led to a deifying of the discipline. Prior to its rise to an institution, advertising was once considered quackery. The use of advertising implied an association of product defect and deficiency. Consumers rarely responded to advertising and certainly did not revere it. Today, advertising, although still detested by some, is a highly regarded institution by many. Its practitioners are exalted and wield significant power and authority. Its icons have become mythological heroes, media sensations, and role models for future generations. There are prized categories for advertising in prestigious award ceremonies, including the Cannes Film Festival and the Grammys. Advertising has produced cherished cultural artifacts and earned esteem through resilience and affiliation. Consequently, it is understandable that millennials desire to be "friends" with brands and products (Jenkins, Ford, & Oliver, 2012).

Through fully comprehending advertising's role and function as a blurred cultural institution, its power is ascertained. Advertising is deeply human and thus subject to the same cycles of development and resilience. Advertising mirrors and shapes culture and lifestyle, much like the people who create it. Throughout its history visionaries that strove to elevate the profession have propelled advertising forward. Despite crisis, patterns suggest that advertising's future may reveal the same. Contemporary visionaries can reinvent the future for advertising.

There is significant power in blurring. However, the power of blurring does not exclude the intermingling and complexity that accompany it. As advertising has ascended to the ranks of comparable institutions such as religion and education, it has become a fundamental component of society and established norms. Moreover, its blurring has established its primacy as a shared institution. Advertising began as a system of the market but today wields the influence of an institution of tradition. It is doubtful that advertising will be easily extricated from its complex relationships without the collapse of major aspects of society.

Awareness, shifting, and problem solving are required to restore productivity in advertising as well as shape its new identity. Neither the continual state of crisis nor the collapse of this institution will be an isolated event. Extraordinary problem solving and inclusion are required to remedy this perplexing dilemma. Due to the impact of triadic convergence, the contemporary advertising environment supplies both the knowledge and power to create significant change.

APPENDICES

The following appendices present summarized details of the framework used to construct this book. They provide information that clarifies analyses and conclusions, and they serve as an easy reference for supportive materials to deepen understandings of the subject matter through theoretical and pragmatic application.

Appendix A: Development of American Advertising 1840s–Present

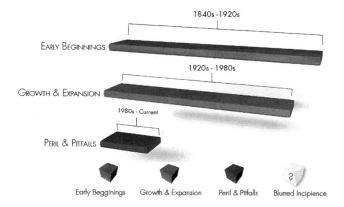

Jenkins offers this model and description to increase understanding surrounding the development of advertising.

- *Early Beginnings* (1840s–1920s). Within this phase of development, advertising knowledge was unable to produce a paradigm that could sustain universal enterprise. Thus, dissention, factions, and concentrations of power characterized this period. Although extraordinarily arduous, this era provided the foundation for future progress and productivity.
- *Growth & Expansion* (1920s–1980s). Within this phase of development, a paradigm was defined that informed the theory and practice of advertising. Subsequently, productivity flourished and advertising was established as discipline. As confidence increased, the profession expanded exponentially. Both scholars and practitioners collaborated to create works that affirmed the paradigm.
- *Perils and Pitfalls* (1980s–Present). Within this phase of development, advertising drifted toward decline. Repeated failure coupled with the inability to sustain productivity revealed the disintegration of the established paradigm. Perils and Pitfalls represents a critical juncture in which advertising can ascend toward revolutionary development through an impending paradigm shift. This period is also closely associated with triadic convergence, a complex phenomenon that elicits unforeseen modifications in worldviews, resulting in blurred practices and ideology.

Please note: the organization for this model is derived from *The Structure of Scientific Revolutions* (Kuhn, 1964).

Appendix B: Triadic Convergence

Jenkins offers this model and description to increase understanding, problem solving, and awareness surrounding triadic convergence and its impact on advertising.

- Triadic convergence is a multifaceted, ever-changing, and complex phenomenon. This dynamic force is comprised of the sophisticated intermingling of its core tenets: media, technology, and culture. Intermingling is characterized by the constant mutative and adaptive synergy achieved between these three forces. This interwoven relationship is shaped and shifts in correspondence with the central locus of power within the triad.

When observing convergence, consideration of the following information is suggested.

1. Advances in technology drive economic processes of cross-ownership.
2. Advances in technology have a direct impact on media.
3. Characteristics of media ownership are normally largely determined by characteristics of technology.
4. Whether or not media ownership is typified by monopoly or competition is normally dictated by the capabilities of the digital environment.
5. Technology and culture have a causal relationship in which the cause-and-effect are sometimes reciprocal and not fixated.
6. Contemporary advertising is characterized by tension generated from the juxtaposition of media monopoly and technological deregulation.

Please note: Observations are derived from the works of de Sola Pool (1983) and Danesi (2012).

Appendix C: Institutional Advertising Model

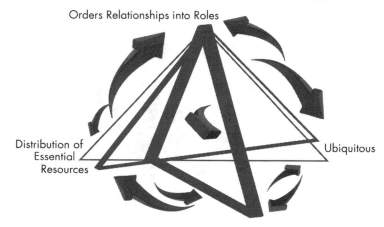

Jenkins has synthesized conceptualizations of advertising, offering this model to increase understanding of institutional views of advertising perspectives and urging the acknowledgement of the following.

- The source of advertising is normally determined by the demands of technology and the location of economic power.
- Contemporary advertising environment is characterized by corporate conglomeration.
- The specific form and nature of advertising messaging is dependent on the economic problem that society recognizes as most pressing.
- The view of society towards consumers, the nature of humanity, and what motivates appropriate market behavior greatly impacts the form and nature of advertising messaging (Carey, 1960).
- The institution of advertising is a fixated, yet evolving, phenomenon.

Please note: Observations are derived from the works of Carey (1960) and Norris (1980).

Appendix D: Impact of Triadic Convergence on Advertising

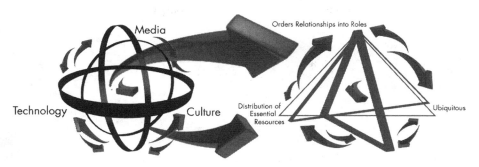

Jenkins offers a model to increase understanding of the impact of convergence on advertising. When observing the impact of convergence on advertising, this investigation urges consideration of the following:

- Advertising is subject to simplified patterns of technological determinism.
- Simplified patterns of technological determinism often fail to take into account technological maturation through lifecycles.
- Advertising will most often behave as a constraint on the direction and pace of change.
- Advertising may appear futile as technology matures.
- Triadic convergence has the potential to create an environment of significant instability within advertising.
- The proliferation of technological deregulation and user-generated content has upset advertising norms.
- A state of crisis has the potential to erupt in environments characterized by corporate monopoly, technological deregulation, and empowered consumer audiences.
- Triadic convergence is a complex phenomenon that has significantly contributed to an impending paradigm shift in advertising.

Please note: Observations are derived from the works of Carey (1960), Norris (1980), de Sola Pool (1983), and Danesi (2012).

Appendix E: Power Disruption Model

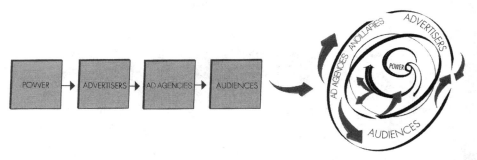

Jenkins offers a model to increase understanding involving the disruption of the power within advertising as a result of the impact of triadic convergence.

- Due to the impact of triadic convergence on advertising, the dominant power structures have shifted.
- Power was once disseminated within a fixed linear structure.
- Power flowed in one direction from advertisers to audiences.
- As a result of triadic convergence, which actively eradicated boundaries surrounding power and significantly contributed to redistribution, power now flows in an erratic circular capacity.
- Power is dispersed in multiple ways and directions among advertisers, mass communication professionals, and audiences.

Appendix F: Internal Disruption Model

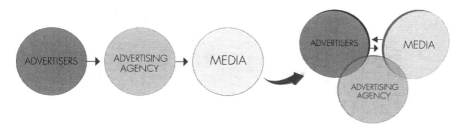

Jenkins offers a model to increase understanding of the disruption of internal relationship structures within advertising as a result of the impact of triadic convergence. This model is particularly useful when examining the internal relationships among advertisers, advertising agencies, and media publishers.

- In the past, a fixed linear relationship existed in which communication primarily flowed directly from advertisers to advertising agencies and then to media publishers.
- As a result of the impact of triadic convergence, fixed internal relationship structures have been disrupted.
- Increasingly, communication occurs directly between advertisers and media publishers.
- Advertising agencies no longer have a fixed communicative role between advertisers and media publishers. This upsets a normative internal relationship structure within advertising.

Appendix G: External Role Disruption Model

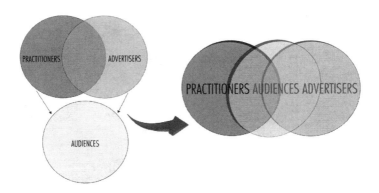

Jenkins offers a model to increase understanding of the disruption of external relationship structures within advertising as a result of the impact of triadic convergence. This model is particularly useful when examining the external relationship structures among advertisers, advertising practitioners, and advertising audiences.

- In the past, a fixed linear relationship existed in which advertising practitioners and advertisers collaborated. Communication primarily flowed directly from advertising practitioners and advertisers. Then messages were delivered to advertising audiences.
- As a result of the impact of triadic convergence, fixed external relationship structures have been disrupted.
- Advertising audiences were once primarily consumers. As a result of the impact of triadic convergence, advertising audiences also have become producers and participatory communities.
- Increasingly, communication among advertising practitioners, advertisers, and audiences does not occur within an ordered hierarchy. This upsets a normative external relationship structure within advertising.
- As a result of the impact of triadic convergence, advertisers, advertising practitioners, and advertising audiences coexist in a collaborative and interactive environment.

REFERENCES

AAAA Study 1970–1974. (1977, October 3). *Advertising Age*.
Ad Age advertising century: Timeline. (1999). *Advertising Age*. Retrieved from http://adage.com/article/special-report-the-advertising-century/ad-age-advertising-century-timeline/143661/
Ad Age encyclopedia of advertising: E-commerce. (2003). *Advertising Age*. Retrieved from http://adage.com/article/adage-encyclopedia/e-commerce/98949/
Adobe. (2007). Adobe: Twenty Five Years of Magic. Retrieved from https://www.adobe.com/aboutadobe/history/
Advertising: A history from 1980–current day. (2009). *Advertising Age*. Retrieved from http://advertisinghistory.blogspot.com/2009_12_01_archive.html
Alkalimat, Y. (2004). Digital literacy: A conceptual framework for survival skills in the digital era. *Journal of Educational Multimedia and Hypermedia, 13*(1), 93–106.
Alka-Seltzer Oral History and Documentation Project, 1953–1987, *Archives Center*, National Museum of American History.
Allen, M. (2012). Targeted sharing: Obama for America. Messina pulls back the curtain on Obama's data mine. *Politico*. Retrieved from http://www.politico.com/playbook/1112/playbook9515.html
Arnheim, R. (1974). *Art and visual perception: A psychology of the creative eye*. Berkeley: University of California Press.
Arnould, E. J., & Thompson, C. J. (2005). Consumer culture theory (CCT): Twenty years of research. *Journal of Consumer Research, 31*(4), 868–882.
Auletta, K. (2009). *Googled: The end of the world as we know it*. New York: Penguin Press.

Bandura, A. (2005). Evolution of social cognitive theory. In K. G. Smith & M. A. Hitt (Eds.), *Great minds in management* (pp. 9–35). Oxford: Oxford University Press.

Banner ads get smarter. (1996). *Advertising Age, 67*(43), 50.

Batchelor, B., & Stoddart, S. (2007). *The 1980s*. Westport, CT: Greenwood.

Belk, R., Guliz, G., & Soren, A. (2003). The fire of desire: A multisited inquiry into consumer passion. *Journal of Consumer Research, 30*, 326–352.

Bell, D. (1976). *The cultural contradictions of capitalism*. New York: Basic.

Benjafield, J. G. (2010). The golden section and American psychology, 1892–1938. *Journal of the History of the Behavioral Sciences, 46*(1), 52–71.

Berelson, B. (1964). The state of mass communication research. In L. A. Dexter & D. M. White (Eds.), *People, society, and mass communications* (pp. 503–509). New York: Free.

Berinito, S. (2010). Four ways of looking at Twitter. *Harvard Business Review*. Retrieved from http://blogs.hbr.org/research/2010/02/visualizing-twitter.html

Bernardin, T., & Kemp-Robertson, P. (2008). Wildfire 2008: Creativity with a human touch. *Journal of Advertising, 37*(4), 131–135.

Bird, A. (2011). "Thomas Kuhn." *Stanford Encyclopedia of Philosophy*. Retrieved from http://plato.stanford.edu/archives/win2011/entries/thomas-kuhn

Brand, S. (1987) *The Media Lab: Inventing the future at MIT*. New York: Viking.

Brandt, R. L. (2011). *One click: Jeff Bezos and the rise of Amazon.com*. New York: Portfolio/Penguin.

Bruell, A. (2013). Publicis Omnicom merger passes antitrust test in U.S. *Advertising Age*. Retrieved from http://adage.com/article/agency-news/publicis-omnicom-merger-passes-antitrust-test-u-s/245074/

Cadman, C., & Halstead, C. (2007). *Michael Jackson: For the record*. Bedfordshire, UK: Authors OnLine.

Calkins, E. E., & Holden, R. (1905). *Modern advertising*. New York: D. Appleton.

Carey, J. W. (1960). Advertising: An institutional approach. In C. H. Sandage & V. Fryburger (Eds.), *The role of advertising in society* (pp. 3–17). Homewood, IL: Richard D. Irwin.

Carey, J. W. (1989). Two Views of Communication: Transmission & Ritual. In J. W. Carey (Ed.), *Communication as culture: Essays on media and society* (pp. 13–36). Boston: Unwin Hyman.

Carlyle, T. (1919). *Past and Present, Book 3: The Modern Worker*. Chapter 1: Phenomena. Oxford: Clarendon.

Carmichael, M. (2010). The great disappearance of wealth. *Advertising Age*. Retrieved from http://adage.com/article/adagestat/great-recession-impacted-affluent-market/146475/

Cawelti, J. G. (2002). Reregionalizing America: A new view of American culture after World War II. *Journal of Popular Culture, 35*(4), 127.

Cherlin, A. J. (2010). Demographic trends in the United States: A review of research in the 2000s. *Journal of Marriage & Family, 72*, 403–419.

Chomsky, N. (1959). A Review of B. F. Skinner's *Verbal Behavior*. *Language, 35*(1), 26–58.

Christodoulides, G., Jevons, C., & Bonhomme, J. (2012). Memo to marketers: Quantitative evidence for change: How user-generated content really affects brands. *Journal of Advertising Research*, 53–64.

Ciochetto, L. (2011). Advertising & value formation: The power of multinational companies. *Current Sociology*, 59(2), 173–85.
Claeys, G. (2000). The "Survival of the Fittest" and the origins of Social Darwinism. *Journal of the History of Ideas*, 61(2), 223.
Cloninger, S. C. (2013). *Theories of personality: Understanding persons: With additional material*. Upper Saddle River, NJ: Pearson.
CNN, (2011). Michael Jackson's 'Thriller' jacket sells for $1.8 million at auction. Retrieved at http://www.cnn.com/2011/SHOWBIZ/Music/06/26/music.memorabilia.auction/
Coleman, C. A. (2010). *Disempowering through definition: A dialogic ethics for understanding consumer vulnerability through Nike's 'Mike and Spike' advertising and African American consumer history*. (Doctoral dissertation, Urbana, University of Illinois).
Coleman, C. E. B. (1998). Advertising: Art as Society's Mirror. *Art Education*, 51, 3, 25.
Colombo, M. (2012). Theoretical perspectives in media-communication research: from linear to discursive models. *Forum Qualitative Sozialforschung/Forum: Qualitative Research*, 5(2), Art. 26, http://nbn-resolving.de/urn:nbn:de:0114-fqs0402261
Craig, R. (1999). Communication theory as a field. *Communication Theory*, 9(2), 119–161.
Creel, G. (1920). *How we advertised America. The first telling of the amazing story of the Committee on Public Information that carried the gospel of Americanism to every corner of the globe*. New York: Harper.
Creswell, J. W. (1994). *Research design: Qualitative and quantitative approaches*. Thousand Oaks, CA: Sage.
Creswell, J. W. (2007). *Qualitative inquiry and research design: Choosing among five approaches* (2nd ed.). Thousand Oaks, CA: Sage.
Crilly, N. (2010). The structure of design revolutions: Kuhnian Paradigm Shifts in creative problem solving. *Design Issues*, 26(1), 54–66.
Curcio, V. (2000). *Chrysler: The life and times of an automotive genius*. Oxford: Oxford University Press.
Dahlen, M., & Edenius, M. (2007). When is advertising, advertising? Comparing responses to non-traditional and traditional advertising media. *Journal of Current Issues and Research in Advertising*, 29(1), 33–42.
Danesi, M. (2012). *Popular culture: Introductory perspectives* (2nd ed.). Lanham, MD: Rowman & Littlefield.
Daunton, M. (1983). The Industrial Revolution: Toil & technology in Britain & America. *History Today*, 33(4), 24.
Delo, C. (2013). Promoted Tweets get big boost in Twitter redesign. *Advertising Age*. Retrieved from http://adage.com/article/digital/promoted-tweets-big-boost-twitter-redesign/245008/
Denzin, N. K., & Lincoln, Y. S. (Eds.). (1999). *The SAGE handbook of qualitative research* (3rd ed.). Thousand Oaks, CA: Sage.
De Sola Pool, I. (1983). *Technologies of freedom*. Cambridge, MA: Belknap.
Deuze, M. (2007). Convergence culture in the creative industries. *International Journal of Cultural Studies*, 10(2), 243–263.
Deuze, M. (2010). *Media life*. Cambridge, UK: Polity.
Dewey, J. (1927). *The public and its problems*. New York: H. Holt and Company.

DiMaggio, P., Hargittai, E., Neuman, W., & Robinson, J. P. (2001). Social implications of the Internet. *Annual Review of Sociology, 27*(1), 307.

Dobson, T., & Willinsky, J. (2009). Digital literacy. *The Cambridge Handbook of Literacy.* Retrieved October 15, 2010, from http://ebooks.cambridge.org/chapter.jsf?bid=CBO9780511609664&cid=CBO9780511609664A026

Donaton, S. (2003) Sting Jaguar deal still serves as model for the music world. *Advertising Age.* Retrieved from http://adage.com/article/viewpoint/sting-jaguar-deal-serves-model-music-world/96186/

Drønen, T. (2006). Scientific revolution and religious conversion: A closer look at Thomas Kuhn's theory of paradigm-shift. *Method & Theory in the Study of Religion, 18*(3), 232–253.

Egri, C., & Ralston, D. (2004). Generation cohorts and personal values: A comparison of China & the United States. *Organization Science, 15*, 210–220.

Enrico, R., & Kornbluth, J. (1986). *The other guy blinked: How Pepsi won the cola wars.* New York: Bantam.

Erevelles, S., Roundtree, R., Zinkhan, G., & Fukawa, N. (2008). The concept of imaginative intensity in advertising. *Journal of Advertising, 37*(4), 131–135.

Feingold, M. (2004). *The Newtonian moment: Isaac Newton and the making of modern culture.* New York: New York Public Library; Oxford University Press.

Fischoff, S. (2005). Media psychology: A personal essay in definition & purview. *Journal of Media Psychology.* Retrieved from http://www.calstatela.edu/faculty/sfischo

Fletcher, H. B. (2012). *Gender and the American temperance movement of the nineteenth century.* London: Routledge.

Foucault, M., Bertani, M., Fontana, A., Ewald, F., & Macey, D. (2003). *Society must be defended: Lectures at the Collège de France, 1975–76.* New York: Picador.

Fox, S. (1984). *The mirror makers: A history of American advertising and its creators.* New York: Morrow.

Fox, S. (1997). *The mirror makers: A history of American advertising and its creators.* Urbana: University of Illinois Press.

Friendly, M. (2009). Milestones in the history of thematic cartography, statistical graphics, and data visualization. National Sciences and Engineering Research Council of Canada. Grant OGP0138748, 1–79.

Frum, D. (2000). *How we got here: The 70's, the decade that brought you modern life for better or worse.* New York: Basic.

Gallagher, C. (2013). Mobile myth-busters: 5 common myths about marketing on mobile devices. *Advertising Age.* Retrieved from http://adage.com/article/digitalnext/5-common-myths-marketing-mobile-devices/245718/

Glass, A. (2007). Understanding generational differences for competitive success. *Industrial and Commercial Training, 39*(2), 98–103.

Godfrey-Smith, P. (2003). *An introduction to the philosophy of science: Theory and reality.* Chicago: University of Chicago Press.

Goldstein, B. (2011). *Cognitive psychology: Connecting mind, research and everyday experiences.* Belmont, CA: Wadsworth.

Gordinier, J. (2008). *X saves the world: How Generation X got the shaft but can still keep everything from sucking*. New York: Viking.

Greenberg, S. (2012). Michael Jackson's "Thriller" at 30: How one album changed the world. *Billboard*. Retrieved from http://www.billboard.com/biz/articles/news/1082851/michael-jacksons-thriller-at-30-how-one-album-changed-the-world

Grimes, G. M. (2011). Paradigm shifts revisited: A deeper fundamental theological engagement with the philosophy of science. *The Heythrop Journal, 52*(2), 181–190.

Gross, D. (2007). *Pop!: Why bubbles are great for the economy*. New York: Collins.

Guertin, C. (2012). Convergence. *Encyclopedia of media and communication*. Toronto: University of Toronto Press.

Gusfield, J. R. (1986). *Symbolic crusade: Status politics and the American temperance movement*. Champaign: University of Illinois Press.

Hall, E. (2010). WPP Acquires Blue State Digital. *Advertising Age*. Retrieved from http://adage.com/article/agency-news/digital-marketing-wpp-acquires-ad-agency-blue-state-digital/147923/

Hampel, S., Heinrich, D., & Campbell, C. (2012, March). Is an advertisement worth the paper it's printed on? The impact of premium print advertising on consumer perceptions. *Journal of Advertising Research*, 41–50.

Hayden, S. (2011). 1984: As good as it gets. *Adweek*. Retrieved from http://www.adweek.com/news/advertising-branding/1984-good-it-gets-125608

Henderson, E. (2009). Philanthropy and big brands need deeper partnerships. *Advertising Age*. Retrieved from http://adage.com/article/goodworks/philanthropy-big-brands-deeper-partnerships/136979/

Hendler, J., & Golbeck, J. (2008). Metcalfe's law, Web 2.0, and the Semantic Web. *Web Semantics, 6*(1), 14–20.

Hendry, L. (2008). Twitter for research: Why and how to do it. *Twitip: Getting More Out of Twitter*. Retrieved from http://www.twitip.com/twitter-for-research-why-and-how-to-do-it-including-case-studies/

Holcomb, J., & Mitchell, A. (2014). The Revenue picture for American journalism and how it is changing. Pew Research Center. http://www.journalism.org/2014/03/26/the-revenue-picture-for-american-journalism-and-how-it-is-changing/

Hovland, R., & Wolburg, J. M. (2010). *Advertising, society, and consumer culture*. Armonk, NY: M. E. Sharpe.

Howe, N., & Strauss, W. (2000). *Millennials rising: The next great generation*. New York: Vintage.

Hoyningen-Huene, P. (1993). *Reconstructing scientific revolutions: Thomas S. Kuhn's philosophy of science*. Chicago: University of Chicago Press.

Hutchins, B. (2011). The acceleration of media sport culture. *Information, Communication & Society, 14*(2), 237–257. doi:10.1080/1369118X.2010.508534

Inukonda, R., & Pereira, D. (2010). Online advertising: The new magic. Presented at the Convergence Culture Consortium, Massachusetts Institute of Technology, Cambridge, MA.

Inwood, J. J. (2011). Geographies of race in the American South: The continuing legacies of Jim Crow segregation. *Southeastern Geographer, 51*(4), 564–577.

Ives, N., Teinowitz, I., Halliday, J., & Steinberg, B. (2008). Media owners resigned to '08 shortfall, brace for tougher '09. *Advertising Age, 79*(35), 3–57.

Jenkins, H. (2006). *Convergence culture: Where old and new media collide.* New York: New York University Press.

Jenkins, J. (2014). Apparitions of the past and obscure vision for the future: Stereotypes of black women and advertising during a paradigm shift. In Goldman, A. Y. (Ed.), *Black women and popular culture: The conversation continues* (pp. 199–223). Lanham, MD: Lexington.

Jenkins, J., Ford, R. L., & Oliver, S. (2012). *Thought leadership: A millennial perspective on diversity & multiculturalism White Paper*, Washington, DC: American Advertising Federation.

Jhally, S. (1997). *The codes of advertising: Fetishism and the political economy of meaning in the consumer society.* New York: Routledge.

Johnson, B. (2011). Old and Improved: Relationships that last for a century. *Advertising Age.* Retrieved from http://adage.com/article/agency-news/ad-agency-marketer-relationships-lasted/148810/

Jordan, V. (1975). Urban league exec charges ad industry neglects Blacks. *Advertising Age.* Retrieved from http://search.proquest.com.proxyhu.wrlc.org/docview/208285448?accountid=11490

Katsanos, C. (2010). Evaluating website navigability: Validation of a tool-based approach through two eye-tracking user studies. *New Review of Hypermedia & Multimedia, 16*(1–2), 195–214.

Keeter, S. (2010). *Millennials: A portrait of generation next.* Pew Research Center. Retrieved at http://pewresearch.org/pubs/1501/millennials-new-survey-generational-personality-upbeat-open-new-ideas-technology-bound

Kennedy, C. (2010). The city of 2050: An age-friendly, vibrant, intergenerational community. *Generations: The Journal of the Western Gerontological Society, 34*(3), 70–75.

Kennedy, L., & Mancini, K. (2006) Boomer segmentation: Eight is enough. *Consumer Insight: Seeing Tomorrow Today.* Retrieved from http://www.acnielsen.com/ci

Kenway, J. (1996). The information superhighway and post-modernity: The social promise and the social price. *Comparative Education, 32*(2), 217–232.

Kern-Foxworth, M. (1994). *Aunt Jemima, Uncle Ben, and Rastus: Blacks in advertising, yesterday, today, and tomorrow.* New York: Greenwood.

Khan, L., & Baig, M. (2007). Cryptanalysis of Keystream reuse in stream ciphered digitized speech using HMM based ASR techniques. *World Congress on Engineering and Computer Science, 10*, 24–26.

Kincheloe, J. L. (2001). Describing the bricolage: Conceptualizing a new rigor in qualitative research. *Qualitative Inquiry, 7*(6), 679–692.

Kolbert, E. (1992, June 15). The 1992 campaign: Media; Clinton campaign to divide its ad work. *The New York Times.* Retrieved from http://www.nytimes.com/1992/06/15/us/the-1992-campaign-media-clinton-campaign-to-divide-its-ad-work.html

Koslow, S., Sasser, S., & Riordan, E. (2003). What is creative to whom and why? Perceptions in advertising agencies. *Journal of Advertising Research, 43*(1), 96–110.

Krugman, D., Leonard, N., Watson, D., & Arnold, B. (1994). *Advertising: Its role in modern marketing.* Fort Worth, TX: Dryden.

Kuhn, T. S. (1964). *The structure of scientific revolutions*. Chicago: University of Chicago Press.
Lafayette, J. (2011). Upfront central: Millennials. *Broadcasting & Cable*, 28.
Lafeber, W. (2002). *Michael Jordan and the new global capitalism*. New York: W. W. Norton.
LaPointe, P. (2011). Marketing on a razor's edge: The need for smarter decisions as the economy goes sideways. *Journal of Advertising Research*, 51(4), 559–560.
Lathan, J. (2006). Technology and reel patriotism in American film advertising of the World War I era. *Film & History*, 36(1), 36–43.
Lee, T., Taylor, R. E., & Chung, W. (2011). Changes in advertising strategies during an economic crisis: An application of Taylor's Six-Segment Message strategy wheel. *Journal of Applied Communication Research*, 39(1), 75–91.
Leedy, P. D. (1997). *Practical research: Planning and design* (6th ed.). Upper Saddle River, NJ: Prentice Hall.
Lindsay, G. (2005). Ad as breakout song launch pad. *Advertising Age*, 26.
Lockwood, L. (2014). The neglected market: Boomers. *Women's Wear Daily*. 4–8.
Lois, G. (2013). Ad campaigns: MTV. Retrieved from http://www.georgelois.com/pages/milestones/mile.mtv.html
Lorigo, L., Haridasan, M., Brynjarsdottir, H., Xia, L., Joachims, T., Gay, G., Granka, L., Pellacini, F., & Pan, B. (2008). Eye tracking and online search: Lessons learned and challenges ahead. *Journal of the American Society for Information Science and Technology*, 59(7), 1041–1052.
Lyle, P. (2007, November). Michael Jackson's monster smash. *The Telegraph*. Retrieved from http://www.telegraph.co.uk/culture/3669538/Michael-Jacksons-monster-smash.html
MacDonald, F., Marsden, M., & Geist, C. (1980). Radio & television studies and American culture. *American Quarterly*, 32(3), 301–317.
Mackaman, D., & Mays, M. (2000). *World War I and the cultures of modernity*. Jackson: University Press of Mississippi.
Marchand, R. (1985). *Advertising the American Dream*. Berkeley: University of California Press.
McCarthy, M., & Wentz, L. (2013). Toyota aligns multicultural marketing under T2 umbrella. *Advertising Age*. Retrieved from http://adage.com/article/news/toyota-embraces-total-market/245736/
McDonough, J. (1998). Pepsi turns 100. *Advertising Age*, 69(29), c1–c5.
McEachern, A. (2012). *Toward a model for industry readiness, preparedness and leadership of African-American millennial media arts students*. (Doctoral dissertation, Howard University, Washington, DC).
McGrath, T. (1996). *MTV: The making of a revolution*. Philadelphia: Running.
McLellan, D. (2008, January). Creative force behind 1980s Pepsi ads. *Los Angeles Times*. Retrieved from http://articles.latimes.com/2008/jan/02/local/me-dusenberry2
McLuhan, M. (1962). *The Gutenberg galaxy: The making of typographic man*. Toronto: University of Toronto Press.
McLuhan, M. (2005). *The medium is the massage*. New York: Gingko.
Miller, G. (2003). The cognitive revolution: A historical perspective. *Trends in Cognitive Sciences*, 7(3), 141–144.
Moriarty, S., Mitchell, N., & Wells, W. (2012). *Advertising & IMC: Principles and practice*. Upper Saddle River, NJ: Prentice Hall.

Moritz, M. (2009). *Return to the little kingdom: How Apple and Steve Jobs changed the world.* New York: Overlook.

Namba, K. (2002). Comparative studies in USA and Japanese advertising during the postwar era. *International Journal of Japanese Sociology*, 11(1), 56–71.

Neff, J. (2013). Who's next to fall? Unilever's massive job cuts put other marketers on notice. *Advertising Age.* Retrieved from http://adage.com/article/news/unilever-s-massive-job-cuts-put-marketers-notice/245587/

Negroponte, N. (1995). *Being digital.* New York: Knopf.

Nicholson, P. (2007, August). Branded for success; McDonald's, others reveal agency world clout when it comes to music. *Billboard*, p. 4.

Nickles, T. (2013). Scientific revolutions: *The Stanford Encyclopedia of Philosophy.* Retrieved from http://plato.stanford.edu/entries/scientific-revolutions/

Nielsen. (2011). State of the media: Social media report. Retrieved from http://www.nielsen.com/us/en/insights/reports/2011/social-media-report-q3.html

Norris, V. (1980). Advertising history-according to the textbooks. *Journal of Advertising*, 9(3), 3–11.

Oates, T. (2009). New media and the repackaging of NFL fandom. *Sociology of Sport Journal*, 26(1), 31–49.

Ohmann, R. (1996). *Selling culture: Magazines, markets and class at the turn of the century.* London: Verso.

Orrell, L. (2009). *Millennials into leadership.* New York: Intelligent Women.

Ortutay, B. (2013). Twitter grabs Super Bowl spotlight. *Associated Press*, Retrieved from http://news.yahoo.com/live-action-twitter-grabs-super-bowl-spotlight-224940470--finance.html

Pasek, J., Kenski, K., Romer, D., & Jamieson, K. (2006). America's youth & community engagement: How use of mass media is related to civic activity & political awareness in 14 to 22-year-olds. *Communication Research*, 33(3), 115–135.

Patton, E. O. (2009). *Under the influence: Tracing the hip-hop generation's impact on brands, sports, & pop culture.* Ithaca, NY: Paramount Market.

Patton, P. (1986, November 9). The Selling of Michael Jordan. *The New York Times Magazine*, 48.

Petough, N. (2010). How to use Twitter for customer service. Business Process Professionals. Retrieved from http://blogs.forrester.com/business_process/2010/01/how-to-use

Pew Research Center. (2014). State of the News Media 2014: Paying for News: The Revenue Picture for American Journalism, and How It Is Changing. Retrieved from http://www.journalism.org/files/2014/03/Revnue-Picture-for-American-Journalism.pdf

Pietila, V. (2005). *On the highway of mass communication studies.* Cresskill, NJ: Hampton.

Pinsdorf, M. K. (1999, December 7). Woodrow Wilson's public relations: Wag the Hun. *Public Relations Review*, 25(3), 309–30.

Pope, D. (2007). Emergence of advertising in America: 1850–1920. *Journal of American History*, 94(2), 660–661.

Potter, P. (1960). The institution of abundance: Advertising. In C. H. Sandage & V. Fryburger (Eds.), *The role of advertising in society* (pp. 18–34). Homewood, IL: Richard D. Irwin.

Raine, G. (2008) S.F. ad man Hal Riney dies. *San Francisco Chronicle.* Retrieved from http://www.sfgate.com/bayarea/article/S-F-ad-man-Hal-Riney-dies-3222241.php#page-2

Raley, R. (2004) eEmpires. *Cultural Critique, 57,* 111–150.

Rao, L. (2011). New Twitter stats: 140M tweets sent per day, 460K accounts created per day. *Tech Crunch.* Retrieved from http://techcrunch.com/2011/03/14/new-twitter-stats-140m-tweets-sent-per-day-460k-accounts-created-per-day/

Raymond, M. (2010). How Tweet it is: library acquires entire Twitter archive. Retrieved from http://blogs.loc.gov/loc/2010/04/how-tweet-it-is-library-acquire

Redniss, J. (2013). Where is social TV headed in 2014? *Advertising Age.* Retrieved from http://adage.com/article/digitalnext/social-tv-heading-2014/245706/

Ries, A., & Ries, L. (2002). *The fall of advertising and the rise of PR.* New York: Harper Business.

Riesman, D., Glazer, N., & Denney, R. (1963). *The lonely crowd: A study of the changing American character.* London: Yale University Press.

Ritchie, A. (2003). *Doing oral history: A practical guide.* New York: Oxford University Press.

Rogers, M. (2012). Contextualizing theories and practices of bricolage research. *The Qualitative Report, 17*(T&L Art., 7), 1–17. Retrieved from http://www.nova.edu/ssss/QR/QR17/rogers.pdf

Rust, R., & Oliver, R. (1994). The death of advertising. *Journal of Advertising, 23*(4), 72–77.

Sacks, D. (2010). The future of advertising. *Fast Company,* 1–28. Retrieved from http://www.fastcompany.com/1702130/future-advertising

Sagan, P., & Leighton, T. (2010). The Internet & the future of news. *Daedalus, 139*(2), 119–125.

Schulman, B. J. (2001). *The seventies: The great shift in American culture, society, and politics.* New York: Free.

Scott, W. D. (1903). *The theory and practice of advertising: A simple exposition of the principles of psychology in their relation to successful advertising.* Boston: Small, Maynard.

Scott, W. D. (1908). *The psychology of advertising in theory and practice.* Boston: Small, Maynard.

Shannon, C. (1948). A mathematical theory of communication. *Bell System Technical Journal, 27,* 379–423.

Shimp, T. A. (2010). *Advertising, promotion, and other aspects of integrated marketing communications.* Mason, OH: South-Western Cengage.

Shiver, J. (1986). "Cosby Show" asking $350,000 and up for ads. *Los Angeles Times.* Retrieved from http://articles.latimes.com/1986-07-01/business/fi-763_1_cosby-show

Short, J., Williams, E., & Christie, B. (1976). *The social psychology of telecommunications.* Hoboken, NJ: John Wiley & Sons.

Simon, P., & Joel, M. (2011). *The age of the platform: How Amazon, Apple, Facebook, and Google have redefined business.* Las Vegas: Motion.

Skinner, B. F. (1957). *Verbal behavior.* New York: Appleton-Century-Crofts.

Slater, W. (2002). Internet history and growth. *Internet Society.* Retrieved from http://www.unc.edu/~tgr/inls572/Slater2002-InternetHistory.pdf

Socolow, M. J. (2004). Psyche and society: Radio advertising and social psychology in America, 1923–1936. *Historical Journal of Film, Radio & Television, 24*(4), 517–534.

Stambor, Z. (2013). Social media advertising grows at the expense of other channels. *Internet Retailer*. Retrieved from http://www.internetretailer.com/2013/01/30/social-media-advertising-grows-expense-other-channels

Starr, P. (2012). An unexpected crisis: The news media in postindustrial democracies. *International Journal of Press/Politics, 17*(2), 234–242.

Stein, J. (1997). How institutions learn: A socio-cognitive perspective. *Journal of Economic Issues, 31*, 729–740.

Stephens, M. (1998). Which communications revolution is it anyway? *Journalism of Mass Communications Quarterly, 75*(1), 9–13.

Stewart, D., Cheng, Y., & Wan, H. (2008). Creative and effective advertising: Balancing spontaneity & discipline. *Journal of Advertising, 37*(4), 131–135.

Stewart, I. (2011). Commandeering time: The ideological status of time in the social Darwinism of Herbert Spencer. *Australian Journal of Politics and History, 57*(3), 389–402.

Stewart, J. (2007). Alan Pottasch, 79; ad exec helped create 'Pepsi Generation' campaign. *Los Angeles Times*. Retrieved from http://articles.latimes.com/2007/aug/02/local/me-pottasch2

Strasser, J. B., & Becklund, L. (1991). Swoosh: The unauthorized story of Nike, and the men who played there. San Diego, CA: Harcourt Brace Jovanovich.

Strasser, S. (2004). *Satisfaction guaranteed: The making of the American mass market*. Washington, DC: Smithsonian Institution.

Stricker, B. (2011). The new workplace currency: It's not just salary anymore: Cisco study highlights new rules for attracting young talent into the workplace. Cisco.com. Retrieved from http://www.cisco.com/c/en/us/solutions/enterprise/connected-world-technology-report/index.html?_ga=1.247273348.1498397356.1415884329

Taraborrelli, J. R. (2009). *Michael Jackson: The magic, the madness, the whole story, 1958–2009*. New York: Grand Central.

Taylor, N., Loiacono, E., & Watson, R. (2008). Alternative scenarios to the banner years: A test of alternative formats to Web banner ads. *Communications of the ACM, 51*(2), 53–58.

Taylor, T. (2009). Advertising & the conquest of culture. *Social Semiotics, 19*(4), 405–425.

Tung, R. (1996). Managing in Asia: Cross-cultural dimensions. In P. Joynt & M. Warner (Eds.), *Managing across cultures: Issues & perspectives* (pp. 233–245). Albany, NY: International Thomson Business.

Tungate, M. (2007). *Ad land: A global history of advertising*. Philadelphia: Kogan Page.

Vansina, J. (1985). *Oral tradition as history*. Madison: University of Wisconsin.

Webster. (2012). Merriam-Webster.com. Retrieved June 2012, from http://www.merriam-webster.com/dictionary

Wernle, B. (2000). Jaguar S-type evokes feeling of style and success. *Automotive News Europe*, p. 18.

Wertheim, J. (2011). Twitter is now a permanent part of the sports firmament. *Sports Illustrated*. Retrieved from http://www.si.com/more-sports/2011/07/14/twitter-thinkpiece

West, D., Kover, A., & Caruana, A. (2008). Practitioner & customer views of advertising creativity. Same concept, different meaning? *Journal of Advertising, 37*(4), 35–45.

Williams, E. (2010). *This is advertising*. London: Lawrence King.

Williams, R. (1980). *Problems in materialism and culture: The magical system*. London: Verso.

Williams, R. (2001). I'm a keeper of information: History-telling and voice. *Oral History Review, 28*(1), 41–63.

Wilson, C. C., Gutiérrez, F., & Chao, L. M. (2003). *Racism, sexism, and the media: The rise of class communication in multicultural America*. Thousand Oaks, CA: Sage.

Wirtz, B. (2001). Reconfiguration of value chains in converging media and communications markets. *Long Range Planning*, (34), 489–506.

Wolf, C. (2013). Big Bang redux. *Advertising Age, 84*(28), 7.

Wood, O. (2012). How emotional tugs trump rational pushes: The time has come to abandon a 100-year-old advertising model. *Journal of Advertising Research*, 31–39.

Yeomans, M. (2010). Social media screw-ups: A brief history. *Advertising Age*. Retrieved from http://adage.com/article/digitalnext/social-media-screw-ups-a-big-missteps/146314/

Young, D. (2005). Sacrifice, consumption, and the American way of life: Advertising and domestic propaganda during World War II. *The Communication Review, 8*(1), 27–52.

Yuan, David. (1998). The celebrity freak: Michael Jackson's "grotesque glory," In R. G. Thomson (Ed.), *Freakery: Cultural spectacles of the extraordinary body* (pp. 368–384). New York: New York University Press.

Zhang, E. (2009). Examining media convergence: Does it converge good journalism, economic synergies, and competitive advantages? (Doctoral dissertation, University of Missouri–Columbia). Retrieved from Proquest. (3371109)

INDEX

A

"A New Beginning" 80
Accumulation of anomalies 133
Acura 103
Adaptive campaigns 160
Adobe 97–99, 157
Adobe Illustrator 97–98
Adobe Photoshop 98
"Ads as news" 19
AdSense 117
Advertising 129–132
    blurring 155
    concentration of power 142
    core disciplines 155–156
    crisis situation 2, 13, 153–154
    defined 129
    development of. *See* Historical chronology
    diversity and inclusion 160
    external disruption model 158–159, 176
    fixed yet continuously evolving phenomenon 6
    in-feed 119
    institution, as 129–130
    institutional advertising model 6–7, 130, 172
    instrument of social control 131
    internal disruption model 157, 175
    malleable state 8
    measurement 113–118
    media, and 131
    monopolies and conglomerates 5
    moving forward 163–167
    power disruption model 7, 163, 174
    quackery, as 167
    self-serve 119
    supporters/detractors 167
    technological deregulation 5–6
    triadic convergence, and 132–134, 157, 173
    ubiquitous nature 130
Advertising agencies 14–16, 103, 113, 116, 157
Advertising anomalies 155
Advertising budgets 153

Advertising Division 31–32
Advertising expenditures 2, 50
Advertising factions 18–21, 120
Advertising industry 129
Advertising measurement 108, 113–118
Advertising professionals 140–141, 165–166
Advertising silos 120
Advertising visionaries 13–18, 167
AdWords 117
Affiliation 166
African American-owned agencies 60
African Americans 24, 60
Age cohorts
 baby boomers 148–150
 millennials 150
Age of Enlightenment 25–26
Agency proponents 20
Air Jordan advertising campaign 87–90
Alka-Seltzer 56
Always Pure 17
Amazon 116–117
American Association of Advertising
 Agencies 21
American Telephone and Telegraph
 Company 23
American temperance movement 21–22
Anomalies 133, 155
Apple 82–83, 97
Apple's "1984" commercial 82–83
Armstrong, Lance 146
Aschberner, Steve 90
Atmosphere advertising 20–21
Audience segmentation 164
Auletta, Ken 118
Autodidacticism 160
Avon 153
Ayer, Francis Wayland 17–18

B

Baby boomers 55, 148–150
Baker, Rick 78
Barton, Bruce 46

Basquiat, Jean-Michel 72
Bates, Charles Austin 21
BBDO 47, 80, 83, 116
"Beat It" 76–77, 81
Bell, Alexander Graham 30
Benatar, Pat 73
Bergin, John 80
Bernbach, William 52–53, 55
Bezier curves 98
Bezos, Jeff 117
"Billie Jean" 75–76
Blue State Digital 114–115
Blurring 155, 167
Bolt, Usain 146
Boost Mobile 103
Bowerman, Bill 84
Bowie, David 74
Brand New Day (Sting) 110
Brand utility 161
Branded entertainment 108–111
Branding 166
Budweiser 103
buildabrand.com 116
Burnett, Leo 52–53
Burrell Communications 160
BuzzFeed 120

C

Cable television 72, 89
Cadillac 103
Calkins, Elmo 38
Cannes Film Festival 167
Capitalism 27
Carey, James 6, 129
Carnegie, Andrew 23
Carphone Warehouse (CPW) 144–145
CCT. *See* Consumer culture theory (CCT)
Cell phones 143
Census data, 148, 150
Centralized/decentralized power 134
Charles Barkley I Am Not a Role Model
 campaign 92

Chiat/Day 82–83, 87, 89
Chicago School of Advertising 53
Chomsky, Noam 56
Chow, Lee 82
Chrysler, Walter 43
Chrysler Corporation 43
Civil Rights Act (1964, 1965) 59
Civil rights era 54
Clinton, Bill 115
Coca-Cola 63, 103, 153
Coe, J. E. 15
Coe, Wetherill & Company 15
Cognitive patterns 71
Cognitive processes 136
Cognitive revolution 56–57
Collectibles 72
Colossal expansion 47–51
Commission system 62
Committee Encouraging Corporate Philanthropy (CECP) 159
Committee on Public Information (CPI) 31
Communal activities 162
Complexity 155, 166
Computer graphics 103
Concentration of power 30, 142
Cone & Belding 62
Conglomerates 2, 142, 154, 164
Conill advertising 160
Consumer apathy 166
Consumer culture theory (CCT) 162
Consumer empowerment 160
Consumerism 71
Contemporary culture 129
Convergence 100, 125–126, 142. *See also* Triadic convergence
Convergence crisis 2
Convergence Crisis, The (Jenkins)
  parts 2
  points of intersection 2
  questions answered 2
  underlying intention 2
Coolidge, Calvin 116
Copeland, Miles 110

Copy testing 71
Cosby, Bill 68
Cosby Show, The (TV) 72
CPI. *See* Committee on Public Information (CPI)
CPW. *See* Carphone Warehouse (CPW)
Crawford, Joan 79
Creationism 26
Creative practitioners 165–166
Creative revolution 51–55, 156
Creativity 140–141, 162, 166
Creel, George 32
Creel Committee 12, 31–32
Crisco shortening 22
Crisis 133–134, 140, 161, 164, 166
Crocker, Charles 23
Crowd sourcing 142
Cultural benchmarks 8
Cultural fusion 42–44
Cultural inferiority 54
Culture 148

D

Dallas (TV) 72
D'Arcy, William C. 31
Darwin, Charles 26
Data visualization 144
DataXu 116
de Sola Pool, Ithiel 5, 100, 125, 128, 139, 141, 171, 173
Decentralized/centralized power 134
Delivery technologies 147
Demand creation 166
Democratized information 97–100
Demographics 148
Denim jeans 64
Deregulation 71, 100, 147
Deutsch 111, 115
Deutsch, Donny 115–116
Development of American advertising. *See* Historical chronology
Dichter, Ernest 48

Digital interruption models 136
Digital volunteerism 115
Digitization 104
Disintegration 155
Dot-com bust 107
Dot-com frenzy 105–106
Douglas Shoes 18
Doyle Dane Bernbach (DDB) 52, 56
Dunn, B. L. 35
Dusenberry, Philip 80, 116
Dynasty (TV) 72

E

eBay 116
Eisenhower, Dwight 49–50
Electronic interdependence 128
Electronification 5, 128, 157
Emotional engagement models 136
Empowerment 158, 160
Ethnic communities 54. *See also* African Americans
Evers, Medgar, assassination of 58
Exemplary professionals 165
External disruption model 158–159, 176
Extraordinary achievement 159

F

Facebook 119
Factions 18–21, 120
Falk, David 86, 88
Fall of Advertising and the Rise of PR, The (Ries/Ries) 63
Fan-voted endings 160
Fashion era 64
Federal Trade Commission (FTC) 59
Fee system 62
Female buying power 148
File sharing 142
FM radio 75
Focus group 71

Ford Motor Company 63
Fox, Michael J. 81
Fragmentation 146
Freud, Sigmund 35

G

Gallup, George 45
Generation X 92
Generational cohorts
　baby boomers 148–150
　millennials 150
Geodemographic clustering systems 63–64
George Creel Committee on Public Information 12, 31–32
Geschke, Charles 97
Gestalt psychology 49
Gilded Age 16
Global conglomerates 2, 154
Global recession 134
Global village 128
Gmail 118
Goodby, Silverstein & Partners 105
Google 117–118
Google+ 118
Google Chrome 118
Google Docs 118
Google News 118
Googled: The End of the World As We Know It (Auletta) 118
Grammy Awards 167
Grassroots movements 120
Great Depression 45
Greenberg, Steven 76

H

Hall & Oates 74
Hannibal, Edward, 55
Haring, Keith 72
Harn, O. C. 32
Harrison, Mike 75

INDEX

Hayden, Steve 82, 83
Hayes, John 109
Heekin, James 61
Historical chronology 9–121
  Adobe 97–99
  advertising agencies 14–16, 103, 113, 116
  Age of Enlightenment 25–26
  baby boomers 55
  branded entertainment 108–111
  Chicago School of Advertising 53
  cognitive revolution 56–57
  colossal expansion 47–51
  commission system 62
  concentration of power 30
  creative revolution 51–55
  cultural fusion 42–44
  democratized information 97–100
  denim jeans 64
  dot-com bust 107
  dot-com frenzy 105–106
  early beginnings (1840s–1920s) 13–39, 170
  factions 18–21, 120
  fashion era 64
  fee system 62
  geodemographic clustering systems 63–64
  Gilded Age 16
  Great Depression 45
  growth & expansion (1920s–1980s) 41–93, 170
  Industrial Revolution 22, 26
  institutional paradigm 33–34
  interdisciplinary collaboration 34–36
  Internet 101–104, 107
  Marlboro Man 50
  measurement 113–118
  MTV 73–75
  national products 17–18
  Nike 83–92
  overview 169–170
  paradox & pitfalls 95–121, 171
  planned obsolescence 38
  presidential election campaigns 113–116
  psychological considerations 34–35, 49, 56
  radio advertising 44–45
  representation of ethnic communities 54
  Roaring Twenties 42–43
  self-indulgence 65
  Shannon-Weaver model 46
  societal concerns 51
  strategic use of modernity 36–39
  television advertising 48–50, 72
  "total agencies" 61–62
  transformative technology 101–104
  turning point (1917) 31–33
  visionaries 13–18
  World War I 31, 33, 37
  World War II 46–47
History 156
Hoagland, Joseph C. 17
Holland, Maury 48
Home Shopping Network 72
Hotwired 101
Houston, Herbert S. 31
"How Dotcoms Killed off the Ad Agencies" 106
Hubbard, Gardiner 30
Hughes, Chris 114–115
"Human kind" 162
Human passion points 162

I

I Want My MTV advertising campaign 73
Ideation 166
IMC. *See* Integrated marketing communications (IMC)
In-feed advertising 119
Inclusion 167
Inclusive cultural markets 160
Incubators for innovation 166
Independent ventures 160
Industrial Revolution 22, 26
Industry of mass communications 140
Infomercials 72
Information 132
Initiative Media Worldwide 104
Innis, Harold 128

Instability 131, 134, 139–140
Institutional advertising model 6–7, 130, 172
Institutional paradigm 33–34
Institutions 139
Integrated marketing communications (IMC) 78, 135–136, 161
Interconnected networks 104
Interdisciplinary collaboration 34–36, 136
Intermingling 126, 140
Internal disruption model 157, 175
Internet 101–104, 107, 128
Internet pricing strategies 112
Interpress 97
Interpublic Group 2, 104, 154
interTrend Communications 160
Ivory Soap 18, 22

J

J. Walter Thompson (JWT) 20, 30, 47–48, 57, 62, 99
Jack Tinker and Partners 56
Jackson, Bo 92
Jackson, Michael 68, 74–81
Jaguar 110
James, Rick 74
Jell-O 68
Jim Crow legislation 24
John Falls Shop 60
Johns, W. H. 31
Jones, L. B. 31
Jordan, Michael 68, 84–90, 92
Jordan, Winthrop 60
"Jordan Flight" commercial 89
Joy, John E. 15
Jung, Carl 35
JWT. *See* J. Walter Thompson (JWT)

K

Kellogg's Corn Flakes 22
Kennedy, John, assassination of 58

Kennedy, Robert, assassination of 58
Kerner Report 54
King, Martin Luther, assassination of 58
Knight, Phil 85
Knowledge and awareness 164–166
Korda, Reva 60
Kraft Foods 48
"Kraft Hands" spots 48
Kraft Music Hall radio program 48
Kraft Television Theater 48
Kroll, Alex 63

L

Laissez-faire economies 27
Landis, Deborah 78
Landis, John 77
Lasker, Albert 15, 20–21, 33
Lasswell, Harold 47
Lauper, Cyndi 74
Lee, Spike 92
Leo Burnett advertising agency 50
Levy's Jewish rye bread 52
Live audience voting system 160
Locke, John 25
Locke's doctrine of natural rights 26
Lois, George 61, 73
Lombroso, Cesare 36
Lord, Daniel 32
Lotame 116

M

MacManus, Theodore, F. 20, 43, 53
Madonna 74, 81
"Magical System, The" (Williams) 112
Making Michael Jackson's Thriller 75–76
Malcolm X, assassination of 58
Market-driven advertising 20
Market saturation 166
Marlboro Man 50
Mass communications 139–140

Mathematical Principles of Natural Philosophy (Newton) 25
McCann, Harry 20, 30
McCann Erickson 30, 47, 49, 57
McCartney, Paul 76
McElroy, Neil 49
McLuhan, Marshall 128
Measurement 108, 113–118
Media conglomerates 142. *See also* Global conglomerates
Media departments 103
MediaMath 116
Mellon, Andrew 23
Messina, Jim 115
Metcalfe's law 104
Michael Jackson Burn Center 81
"Michael Jackson's Thriller" 75
Migratory consumer behavior 146, 166
Millennials 150, 167
Mindshare 104
Mobile devices 143, 161
Modern Advertising (Calkins/Holden) 37
Modern communication 140
Moore, Peter 85, 87, 90
Moore's law 104
Morgan, J. P. 23
"Morning in America" commercials 116
MTV 73–75
Music commodification 111
Music Television (MTV) 73–75

N

N. W. Ayer & Son 15, 22–23, 30
Napster file-sharing controversy 109
National products 17–18
Natural right of property 26
Natural selection 26
Negroponte, Nicholas 104, 125, 128, 135
New media ecosystem
    creativity 140–141
    defined 139
    institutions 139

mass communications 139–140
research 140
self-efficacy 141
New York Telephone Company 30
Newton, Isaac 25
Newtonian World Machine 25
Newton's laws of motion 25
Ney, Edward 61
Nicholson, Peter 111
Nicks, Stevie 74
Nielsen, A. C. 45
Nielsen rating system 49
Nike 68, 83–92
Nineteen Eighty-Four (Orwell) 82
"1984" commercial 82–83
Nixon, Richard M. 58, 116
Norris, Vincent 6, 129

O

Obama, Barack 113
Obsoletism 38
Ogilvy, David 52–53, 60
Omnicom Group 2, 99, 104, 154
On the Origin of Species (Darwin) 26
O'Neal, Shaquille 146
Overlapping phenomena 132–134

P

Paley, William 125
Palmer, Volney 13–16, 18
Paradigm 133
Paradigm shift 132–133, 154–156
Paraphernalia 72
Participatory audiences 160
Participatory culture 147–151
Partnership for a Drug-Free America 80
Patron Saint of Advertising 49
Penalty of Leadership 20
People of color 54, 148. *See also* African Americans

Pepsi 68, 79–81, 153
Pepsi Generation 79–81
Personalization 65
Peters, Michael 77
Pettengill, Samuel 16
Philosophiae Naturalis Principia Mathematica (Newton) 25
Pinterest 119–120
PlanetRx.com 105
Planned obsolescence 38
Polar Bears campaign (Coca-Cola) 103
Pop culture milestones 8
Postings 142
PostScript 97
Pottasch, Alan 79–80
Power disruption model 7, 163, 174
Power redistribution 164
Power structures 156
Powers, John E. 19–20
Precision-driven research 71
Premium print advertising 135
Presidential election campaigns 113–116
Price, Vincent 78
Principles of Biology (Spencer) 26
Problem solving 167
Procter & Gamble 63, 153
Professionals 140–141, 165–166
Psychology 34–35, 49, 56, 107
Psychology of Advertising, The (Scott) 35
Psychology of Advertising in Theory and Practice, The (Scott) 23
Publicis Groupe 2, 154, 160

Q

Quinn, Stanley 48
QVC 72

R

Rabinowitz, Josh 111
Radio advertising 44–45

Rand, Paul 53
Ray, Ola 77
Reagan, Ronald 80, 116
Reeves, Rosser 50, 55, 70
Reform era 22
Research 140
Resor, Stanley 35–36
Retail advertising 19
Richie, Lionel 81
Ries, Al 63
Rights of the disadvantaged 24
Riney, Hal 70
Roaring Twenties 42–43
Robinson, Charles 85
Rock superstars 74
Rockefeller, John D. 23, 30
Roe v. Wade 59
Rowell, George P. 16–18
Rowell's American Newspaper Directory 17
Royal Baking Powder 17
Rubicam, Raymond 46

S

Saatchi & Saatchi 160
Saint Bernardino of Siena 49
Sanders, Thomas 30
Sapolio soap 18
Scholarship 140
Scott, Ridley 82
Scott, Walter 23, 35
Seamless communications 161
Segmentation 164
Self-efficacy 141
Self-indulgence 65
Self-regulating mechanisms 27
Self-serve advertising platform (Twitter) 119
Sexual revolution 54
Shannon-Weaver model 46
Sharpe, W. W. 15
She's Gotta Have It (film) 92
Shop TV Canada 72
Short message language systems 102

Small creative boutiques 103
Social contract 159
Social media 119, 142–146
Social presence theory 143, 145
"Speedy," 56
Spencer, Herbert 26
Spike and Mike commercials 92
Sports 145–146
Standard Oil Company 23, 30
Stanford, Leland 23
Starch, Daniel 45
State of crisis 133–134, 161, 164
Stern, David 89
Sting (rock star) 74, 110
Story building 141
Storytelling 141
Strasser, Rob 88, 91
Strategic targeting 21
Structure of Scientific Revolutions, The (Kuhn) 170
Subordinate positioning 54
Sun Belt 71
Super Bowl 2013 146
Survival of the fittest 26
Swedien, Bruce 75, 79

T

Tacit knowledge 165
Target audiences 2, 154
Technological breakthroughs 8, 13, 146–147
Technologically adept companies 161
Technologies of Freedom (de Sola Pool) 100, 125
Technology-driven measurement practices 116
Television advertising 48–50, 72, 135
Temperance movement 21–22
The Bo Knows campaign 92
The Cars 74
The Choice of a New Generation advertising campaign 80–81

"The Girl Is Mine" 76
Theory of Advertising, The (Scott) 23
Thomas, Brent 82
Thompson, James Walter 15
Thorn, Rod 88
Thriller 68, 74–76, 80
"Thriller" 75–79, 81
Tone, Charles 103
"Total agencies" 61–62
Townshend, Peter 73
Toyota Motor Company 160
Traffic departments 48
Triadic convergence 125–129
    commercialization of Internet 158
    consumer empowerment 160
    defined 126–127, 171
    diversity and inclusion 160
    electronification 5, 128, 157
    extraordinary achievement 159, 162
    factors to consider 171
    graphical representation 127
    impact of, on advertising 132–134, 173
    inclusive cultural markets 160
    instability, and 131, 134
    participatory audiences 160
    revolutionary phenomenon 159
    seamless communications 161
    signifier of change 139
    underlying factors 141
Triadic convergence-based advertising strategies 161
Triadic convergence model 4–5, 127, 171
Trout, Jack 63
Tuqiri, Lote 146
Twitter 119, 143–146

U

Uneeda 15
Unilever 153
Universal truths 162
User-generated content 120, 136

V

Value exchange 121
Van Halen, Eddie 74, 76
Vanderbilt, Cornelius 23
Video jockey (VJ) 73
Vietnam War 54
Visionaries 13–18, 167
Volney B. Palmer & Company 15
Voting Rights Act (1968) 59

W

Wanamaker, John 19
War Advertising Council 46
Ward, Artemas 18
Warhol, Andy 72
Warnock, John 97
Watergate scandal 58
Watson, John B. 35
Web 2.0 142
Wells, Mary 60
Whassup! campaign 103
Where You At? This advertising campaign 103
Wieden, Dan 91
Wieden & Kennedy 87, 91–92
Williams, Raymond 112
Williams, Serena 146
Women's suffrage movement 24
World War I 31, 33, 37
World War II 46–47
Worldwide 104
WPP Group 2, 99, 104, 115, 154

Y

Yetnikoff, Walter 74
Young & Rubicam (Y&R) 47, 57, 62–63
Young advertising professionals 166
Youth movement 54

Z

Zappos 145
Zebra Associates 60
Zenith Media 104, 160
ZZ Top 74